SpringerBriefs in Business

SpringerBriefs present concise summaries of cutting-edge research and practical applications across a wide spectrum of fields. Featuring compact volumes of 50 to 125 pages, the series covers a range of content from professional to academic. Typical topics might include:

- A timely report of state-of-the art analytical techniques
- A bridge between new research results, as published in journal articles, and a contextual literature review
- A snapshot of a hot or emerging topic
- An in-depth case study or clinical example
- A presentation of core concepts that students must understand in order to make independent contributions

SpringerBriefs in Business showcase emerging theory, empirical research, and practical application in management, finance, entrepreneurship, marketing, operations research, and related fields, from a global author community.

Briefs are characterized by fast, global electronic dissemination, standard publishing contracts, standardized manuscript preparation and formatting guidelines, and expedited production schedules.

More information about this series at https://link.springer.com/bookseries/8860

Kaan Y. Ciftci • Alex Michel • Patrick Siegfried

The Potential Impact of E-Mobility on the Automotive Value Chain

Springer

Kaan Y. Ciftci
ISM International
School of Management GmbH
Frankfurt am Main, Germany

Alex Michel
ISM International
School of Management GmbH
Frankfurt am Main, Germany

Patrick Siegfried
International Management
ISM International
School of Management GmbH
Frankfurt am Main, Germany

ISSN 2191-5482 ISSN 2191-5490 (electronic)
SpringerBriefs in Business
ISBN 978-3-030-95598-4 ISBN 978-3-030-95599-1 (eBook)
https://doi.org/10.1007/978-3-030-95599-1

This Springer imprint is published by the registered company Springer Nature Switzerland AG
The registered company address is: Gewerbestrasse 11, 6330 Cham, Switzerland

Preface

This book provides an extensive insight about the impact of electro-mobility on traditional automobile manufacturers. Therefore, the research analyzes the drivers of electro-mobility and develops a forecast model with the help of exclusive industry reports from leading investment banks.

The aim of this book is to find out the impact of electro-mobility on the automotive value chain. Expert interviews with the leading automobile supplier Continental, the "big 4" consultancy firm KPMG, the market-leading leasing company Deutsche Leasing, and a VW-Audi car dealer have been executed to gain an insight industry perspective.

In order to provide an in-depth analysis of the automobile industry, electro-mobility cannot be seen as an isolated trend. Thus, the impact of accompanying trends like shared economy, autonomous driving, artificial intelligence, and digitalization will be analyzed in this research, as well.

The research results show that the development of e-mobility is contributed by technological innovations, charging infrastructure, and changing customer preferences. The global electric vehicle sales will surpass the sales of conventional internal combustion engines between 2030 and 2040. In this pace of change, new entrants from cross-industries like technology companies, mobility providers, specialized suppliers, and energy providers are implementing new products and services to the automobile market. This forces traditional automobile manufacturers to adapt their business models and to increase vertical integration. Furthermore, they will integrate new competencies like after sales and financial services, mobility services, and charging infrastructure into their value chain. Due to these changes, manufacturers are moving away from their traditional core business and have to reposition themselves within the automobile industry.

The qualitative research is addressing recent game-changing trends in the automobile industry, of which the future outlook is still shaped by uncertainties. This book will provide a clarifying overview about important indicators, the development and the impact of e-mobility and will provide potential strategies for traditional automobile manufacturers.

Frankfurt am Main, Germany Kaan Y. Ciftci
Frankfurt am Main, Germany Alex Michel
Frankfurt am Main, Germany Patrick Siegfried

Contents

Abbreviations

BEV	Battery electric vehicle
CO_2	Carbon dioxide
E-mobility	Electro-mobility
EV(s)	Electric vehicle(s)
HEV	Hybrid electric vehicle
ICE	Internal combustion engine
MHV	Mild hybrid vehicles
OEM(s)	Original equipment manufacturer
PHEV	Plug-in hybrid
PQ	Pivotal question

List of Figures

List of Tables

Chapter 1
Introduction to the Potential Impact of E-Mobility on the Automotive Value Chain

1.1 Introduction and Objective

If I had asked people what they wanted, they would have said faster horses. (Henry Ford)

Technological development and social responsibility are changing the world's basic idea of mobility and transportation. The world's desire for a sustainable planet and new technological innovations are shifting behavioral thinking of people around the world.

The traditional internal combustion engine (ICE) dominated private transport for about 100 years (California Air Resources Board, 2012). Currently a lot of indicators and industry experts show that electric vehicles (EVs) will replace ICEs. EVs already have been implemented around 1990 yet failed due to the low capacity of battery and the cost of ownership. This time, EVs have a better starting point based on two different factors. The first factor is that e-mobility today has multiple drivers such as technological innovations, urbanization, and policy makers. The second factor is that e-mobility can no longer be seen isolated, because it is coming with other mega trends. Both factors give EVs the potential to become a dominating mass market product in the future.

Besides e-mobility, mega trends like self-driving cars, artificial intelligence, and shared economy are impacting the automotive industry.

The total revenue of the global automobile industry currently accounts 3.5 trillion dollars. According to recent study results from McKinsey, the global automobile industry will potentially increase to more than 7 trillion dollars by 2030. This outstanding doubling is mainly driven by e-mobility, shared mobility, and digitalization (McKinsey, 2016, pp. 10–11).

The traditional idea of getting from point A to B has changed. When it was the goal to reach a destination, it was a matter of effectiveness. Now it is a matter of efficiency. Effectiveness is about how something can be done in order to achieve a specific goal or result. Efficiency is about achieving a goal in the most optimal way and puts the effort into relation with the outcome. In the past a destination could be

K. Y. Ciftci et al., *The Potential Impact of E-Mobility on the Automotive Value Chain*, SpringerBriefs in Business, https://doi.org/10.1007/978-3-030-95599-1_1

reached by means of public transport or private transport. In order to use private transport, it was necessary to own a private vehicle. Today this is not the case anymore. Mobility opportunities these days are much broader and flexible due to EVs and new ride-sharing concepts. It is about which opportunity meets customer preferences the best and which opportunity provides the best input output relation.

As a result, new entrants are accompanying these trends and challenge traditional car manufacturers. Cross-industry companies are entering the automobile industry, adding new products and services to the market. This forces traditional car manufacturers to adapt their value chain and to walk away from their traditional core business.

In order to go along with the trend of e-mobility, it is essential for car manufacturers and suppliers to assess the pace of change in which mega trends like e-mobility will enter the markets. Knowing the development of e-mobility will help car companies to create new strategies and interim solutions (e.g., hybrid engines) for the new era of electric vehicles. Assessing the development of e-mobility allows car manufacturers to schedule their activities and measures, which are required for the change process.

Therefore, this book aims at investigating on the arising question of how e-mobility is impacting the automotive value chain of traditional manufacturers.

The central question of this book will be worked out with the help of expert interviews of the car supplier Continental, the "big 4" consultancy KPMG, the leading German leasing firm Deutsche Leasing, and a German based Audi/ Volkswagen car dealer. The interviews shall give insights about the impact of e-mobility on major automobile manufacturers. Furthermore, it gives answers on how car manufacturers can develop potential strategies for the new era of mobility. Additionally, exclusive research reports from leading investment banks Morgan Stanley, J.P. Morgan, Bank of America Merrill Lynch, and UBS will be used and have been provided for the use of this book. They shall give insights and ensure that conclusions shall be drawn, which contribute to analyze the progress and potential impact of e-mobility.

1.2 Research Design

This research assumes that e-mobility will establish in the automobile market and replaces internal combustion engines within the future.

The objective of the research will be worked out by means of four pivotal questions (PQ).

The main objective of the research is it, to provide a deeper understanding of the biggest changes in the automotive value chain, due to e-mobility. The first central question, therefore, is "how does e-mobility impact the automotive value chain of traditional automobile manufacturers?".

In order to get closer to the answer of the central question, the influencing factors of e-mobility will be identified in PQ 1. This information will provide a firm grasp of how the future development curve of e-mobility will look like.

Pivotal question 1: What are the drivers (influencing factors) of e-mobility?

Based on PQ1, exclusive reports from Morgan Stanley, UBS, Bank of America Merrill Lynch, and J.P. Morgan will be critically analyzed and comprehensibly involved into the scenario analysis in PQ 2. Additionally, consulting reports form Deloitte will be used in order to answer PQ 2. Based on this data an average value will be determined in order to create a development curve for this research.

Pivotal question 2: How the progress curve of e-mobility will look like?

Independently from PQ 1 and PQ 2, it will be analyzed which differences will arise for OEMs due to e-mobility. In order to identify what will change for them due to e-mobility, personal interviews with the head of business development of Continental, partner of KMPG, managers from the leading German leasing company and car dealers will be executed. This will be further supported by online research, reports, and scientific literature.

Pivotal question 3: What will change in large OEMs due to e-mobility?

Based on in which pace e-mobility will progress (LQ 2) and which changes will occur for OEMs (PQ 3), PQ 4 will answer which challenges and opportunities will arise for OEMs. Which other changes will occur and which factors are important for OEMs to consider for the transition. This question will include the pre identified information from PQ 1–3 and therefore will answer the central question.

Pivotal question 4: Which challenges and opportunities does e-mobility offer for OEMs? Which potential strategies can be derived, in order to overcome challenges and realize opportunities?

References

California Air Resources Board. (2012). *Climate Change programs*. Retrieved September 20, 2021, from California Air Resources Board Website: https://www.arb.ca.gov/cc/ab32/ab32.htm

Henry Ford. (n.d.). *Henry Ford, innovation, and that "faster horse" quote*. Retrieved from Havard Business Review: https://hbr.org/2011/08/henry-ford-never-said-the-fast

McKinsey & Company. (2016, February). *Beyond the supercyle*. Retrieved September 20, 2021, from McKinsey & Company Website: https://www.mckinsey.de/files/170216_pm_mgi_beyond_the_supercycle.pdf

Chapter 2
Drivers of E-Mobility

2.1 Climate Change

The impact of global warming is very conspicuous. Due to human influence the earth climate on earth changes. For about 80 years, the surface of land and ocean is increasingly getting warmer, global sea level increases, and global greenhouse emissions are rising (Intergovernmental Panel on Climate Change, 2014).

The driver of climate change is the continuous growing population and economy, which were also the triggers for the industrial revolution between the seventeenth and early nineteenth century (Mokyr & Strotz, 1998).

Numerous studies and scientists have proven that population will continue to grow within the next 100 years. Between 2020 and 2025, the forecasted population will reach 8 billion, and, between 2050 and 2060, the forecasted population reaches 10 billion (Roser, 2018).

Analogous to the growing population the carbon dioxide concentration within the earth's atmosphere is increasing as well.

In 2014, the concentration of carbon dioxide accounted 398 parts per million. In 1750, this figure only accounted 278 parts per million (Vaughan, 2015). Without having significant measure for this progress, carbon dioxide concentration will increase further and climate change will drastically impact the climate system. A heating planet, melting ice, rising sea level, environmental disasters, and heavy storms are the consequences. This will lead to worsening living conditions where people have to escape from their living spaces. According to a study from Greenpeace, there are already 20 million refugees who are currently escaping from their living place due to climate change (Greenpeace, 2014).

In conclusion, climate change has significant consequences and will demolish the earth and the economy with it, if there will not be any measures taken against it. The fear of this future state increased the awareness of population during recent years. Governments, organization, and companies already began to fight back climate change by, e.g., global climate change regulations and agreements. However,

K. Y. Ciftci et al., *The Potential Impact of E-Mobility on the Automotive Value Chain*, SpringerBriefs in Business, https://doi.org/10.1007/978-3-030-95599-1_2

nations and industries are seeking for possible measures which have the power to stop the climate change by reducing carbon dioxide in a long-term and sustainable view.

The world's industry society is the main polluter who is responsible for global warming. Thus, it needs to be looked for new ideas which have the potential to transform many industries and not just some sectors. Digitalization is one form of transforming industries and pretends the earth from the consequences of climate change. Recently digital technologies are increasingly developed in almost every big industry and have a huge potential for fighting against climate change.

The automotive industry is considered to be one of the "big industries" and is one key industry regarding the fight against global warming (Lempp & Siegfried, 2021). Private and commercial cars make up approx. 25% of the total CO_2 emissions of the world (Blume, 2015). With total revenue of 3.5 trillion dollar, the global automotive industry has produced 94 million cars and commercial vehicles in 2016 and this amount is likely to increase within the future due to growing population and consumption (Funda, 2014; International Organization of Motor Vehicle Manufacturers, 2018).

Within the automotive industry digital technologies are already introduced and have caused three major trends which are electro-mobility, autonomous driving, and shared economy/connectivity (Bitkom e.V., 2015).

Every trend comes with digitalization and has cross-industry-potential to fight against climate change.

In general there are three possible ways of reducing CO_2 emissions of the automobile industry which are all covered by the three major trends.

1. Alternative powertrains
2. The reduction of the total amount of cars
3. Efficiency of conventional cars (ICEs)

The biggest potential of reducing CO_2 within the car industries are alternative powertrains. The electric engine is the best developed alternative powertrain compared to other alternative engines like, e.g., fuel cells or water driven engines. EVs, therefore, have the most potential to be the game-changing way of transforming a key industry for climate change (Siegfried & Strak, 2021).

Additionally, less cars are producing less CO_2. Therefore, new mobility concepts like car sharing have been introduced already. Shared economy is causing that the principal idea of ownership is changing into sharing cars with other people. Shared economy can reduce the total amount of cars in the world and also reduce traffic and parking spaces within urban cities.

In order to get conventional cars more efficient car manufacturers have to reduce the fuel consumption of the ICE. This can be achieved, e.g., by using lightweight materials or by developing compressor systems of the ICE.

Digitalization and e-mobility are both products of the rising awareness for climate change. The importance of e-mobility, shared economy, and autonomous driving for climate change is humongous. E-mobility can impact whole industries and reduce the total amount of CO_2 on the long-term run. Thus, global climate change is one big driver of e-mobility.

2.2 Historical Background and Technological Leaps

The idea of electric vehicles already existed when the first cars were manufactured. So the idea of EV is not a new invention. In 1912, EVs have been already produced in the USA (Maxwill, 2012).

The French physicist Gaston Plante already invented the rechargeable battery.

At the electricity fair in Paris in 1881, the Frenchman Gustave Trouvé showed an electric vehicle with an electric motor with a lead-acid battery. It is today considered as the first electric vehicle in the world.

From 1896 to 1912, EV had their peak time, and by that time electric vehicles had a range of more than 100 km. Around 1900, the share of electric vehicles accounted 38%. Around 34,000 electric vehicles registered in the USA with over 500 brands of electric cars worldwide (Desmond, 2016).

EVs prevailed over the ICEs for almost a century and finally dominated the market. But around 1911, EVs lost their strong position in the market.

In 1911, the American engineer Charles F. Kettering invented the electric starter for the gasoline engine, which eliminated the power-consuming crank. The number of gas stations increased dramatically and oil prices were low. Additionally, the advertisement for the ICEs caused an image change. The market share of EVs decreased rapidly until they reached the end of production around 1920.

Although the vehicles had a limited range compared to ICEs, the benefits of EVs were their smoothness compared to ICEs, there were no exhaust gas, and EVs were easy to use (Siegfried & Strak, 2021).

Until the beginning of 2000, EV considered to be a niche product. This was changed by the oil crisis in 1990. This provided the initial ignition return of EVs. The CARB (California Air Resources Board), for example, introduced the law for the commission offering for EVs, which led to strong developments of automobile manufacturers producing EVs (California Air Resources Board, 2012).

Car manufacturers were still struggling with the development of electric vehicles, but since 2003, a strong development of EVs has been identified. Small independent companies produced EVs.

In the further course, Tesla Motors set new standards in e-mobility. In 2006, Tesla started to sell the Tesla Roadster as their first series product. The Roadster then had a range of up to 350 km (Conrady & Buck, 2012).

It is obvious that since 1881, the world has experienced huge fundamental changes. However, the question arises as of which specific technological changes have contributed to the introduction of EVs, today (Rohrer, 2017).

The most obvious change is the development of battery for the automotive application. The first batteries were lead-acid batteries. In 1990, lithium-based ion batteries have been invented for the use of EVs. The invention of lithium ion bases batteries was the trigger which contributed to the success of today's EV development.

In order to realize why lithium ion batteries were an industry revolution, the two factors of batteries have to be understood. There are two crucial factors for batteries that have to be considered for identifying which specific change led to the reinvention of today's EVs. The first factor is the energy capacity and the second factor is the weight. By putting these two factors into relation, the lead-acid battery and the lithium ion battery can be compared.

The lead-acid battery requires a weight of 30 kg in order to store 1 kWh electricity.

The lithium ion battery requires only 6 kg in order to store 1 kWh electricity.

By looking at a practical example to the new models of Tesla, 1 kWh requires approx. 7 kg (600 kg battery/85 kWh) (Teslarati Network, 2013).

This shows that lithium-ion technology provides the basis for the future EVs. Based on lithium-ion technology, the range of EV increases significantly, while the weight is decreasing. As a result, the potential of performance improvements for EVs such as acceleration increases. Additionally, the lithium-ion technology eliminates the risk of getting in contact with dangerous chemicals. Lead-acid batteries included very caustic and dangerous chemicals. Furthermore, lithium-ion technology shows a higher life cycle and can be charged more efficiently than lead-acid batteries which can only be charged for 100–150 times (D'Aprile et al., 2016).

In conclusion, lithium-ion batteries have a lighter weight, increased driving range, are much more consumer friendly, and show a higher lifetime. These specific changes of the battery have led to the progress of today's EVs. Due to this technology, car manufacturers like Tesla have specialized on further developing of lithium-ion batteries. Extended range, faster charging, and better performance due to this technology have led to an improving image of EVs toward the automotive industry (Stapleton, 2013).

Not only technological leaps of battery but also the technology of electric infrastructure has improved significantly. Technological milestones like smart chargers or vehicle to grid provide the basis for infrastructure solutions of the future. These technologies do not focus on the first problem which dealt with the installation of charring stations, but go further. These technologies focus on the efficient allocation of energy within the infrastructure cycle. This is a fundamental element in order to successfully use EV in a long-term view (Ernst & Young Global Automotive Center, 2012).

Summarized, the historical progress of technology and technological leaps enabled manufacturers to build EVs which are competitive enough for the global automotive market. These technological leaps have formed the foundation on which todays EVs are constructed. However, the technology for EVs just started and has to be elaborated strongly within the future. EVs are highly shaped by technological progress. Range, weight, performance, battery, and infrastructure can only be improved by technological progress. This is the reason why technology is a strong driver of e-mobility.

2.3 Political Regulations

In order to push the progress of e-mobility, governments can set specific regulations, which can address numerous issues. Regulations can address the efficiency of new registered cars, mileage and way of driving (driving-style), existing vehicle fleet, energy source, and infrastructure.

Since 2009, the EU agreed on CO_2 regulations in order to reduce CO_2 emissions until 2015.

Specifically, the EU Regulation EC 443/2009 set a CO_2 limit for passenger vehicles of 130 g/km CO_2, which should be achieved by 2015.

In the USA the limit for car CO_2 emissions is at 157 g CO_2/km and in Japan 141 g CO_2/km (Deutsche Bank, 2011).

Looking at the past decades, it can be observed that regulations and CO_2 restrictions are increasingly tightening in the USA, Europe, and Asia. The EU has decided to tighten the CO_2 limit from 130 g CO_2/km to 95 g CO_2 /km until 2020. The European Commission is currently preparing a regulatory proposal for the period after 2020, which will be even tougher. The USA targets a CO_2 121 g/km, China 117 g/km, and Japan 105 g/km until 2020. So the EU has the most ambitious CO_2 target plan to achieve (Verband der Automobilindustrie, 2015).

Nevertheless, the question arises if these regulations really have an impact on the progress of EV or are they just increasing the pressure on OEMs in the automotive industry, which can lead to industry crises like the VW emission manipulation scandal in 2014 (Hotten, 2015).

The general purpose of these regulations is to reduce car emissions in order to effectively protect global climate damage due to the greenhouse effect.

The existing regulations in the USA, Europe, and Asia are mainly addressing OEMs and force them to produce more efficient cars. By setting stricter targets, OEMs have to make their internal combustion engines more efficient or they have to produce cars with alternative powertrains such as electric engines. Due to the fact that the CO_2 targets are getting tougher every year, OEMs must include EVs into their product portfolio, because otherwise the targets cannot be achieved in long term. As a result, the share of EVs should increase.

In fact CO_2 emissions of passenger cars are decreasing and OEMs are increasingly producing more efficient conventional cars. However, the share of EVs is not increasing significantly (European Commission, 2017).

There are two major reasons for that. First, the existing regulations are highly focused on having more efficient cars. But by setting stricter targets, OEMs have to make efficient cars which are more expensive than old environmental unfriendly cars, consequently customers are sticking to their old cars and will not buy a new expensive car. Consequently, the policy makers are not addressing the existing vehicle fleet with their regulation.

Secondly, in order to achieve CO_2 targets, alternatively OEMs could produce and sell more EVs instead. The problem hereby is also that the demand for EVs is not high enough, so that OEMs are struggling with the sales of EVs. EVs require a reliable infrastructure, which is not considered by the regulations of the policy makers. Consequently, the demand for EVs is not increasing.

On the one hand, the regulations are not addressing the trend of customer preferences which are sticking to their old environmental unfriendly car, and, on the other hand, the requirements for successfully introducing EVs are not sufficiently supported by the existing regulations. As a result, the share of EVs is not increasing sufficiently.

Despite that the EU policy makers are increasingly setting commissions and quotes for eco-friendly car customers (T-Online, 2017).

Additionally, quotes for EVs have been already introduced in China and are planned also for European countries (Giesen et al., 2017).

The EU policy makers are already working on new regulations after 2020 which are not only including a product based regulation, but also focus on customer preferences (demand side), infrastructure (charging stations, etc.), digital connectivity of EVs, and on the renewal of the existing car fleet (Wilkens, 2017).

The German Association of the Automobile Industry (Verband der Automobilindustrie, 2015) provides a very broad regulation proposal which addresses the relevant issues to effectively reduce CO_2 emissions and increase EV share.

According to the VDA (Verband der Automobilindustrie, 2017) the ideal regulation after 2020 should still include the basic principles of CO_2 limitation. Additionally, effective CO_2 emission reduction should not only be addressed to the supply side (OEMs) with producing new more efficient cars, but also should address a broader field.

They propose to set regulations which further improve the efficiency of their new vehicles with, e.g., reduction mandates, as long as they increase the share of electric engines. Technological innovations which reduce CO_2 emissions should also be considered in the future.

On the other hand, the previous supply-oriented regulation has to be added with an effective strategy for the demand side. The purpose is to reduce CO_2 emissions of existing vehicles, not only from new registered cars. Additionally, intelligent digital technologies for traffic can be addressed. In order to actively increase the share for EVs, a demand-based policy by EU policy which supports electric engines, is required (Verband der Automobilindustrie, 2017).

As a conclusion it can be observed that the future regulations will address more supply side as well as demand side issues which are passively and actively supporting market growth of EVs. Therefore, the progress of EVs is expected to rise by policy making regulations in the future.

2.4 Energy Consumption and Urbanization

The awareness of sustainable energy is changing the world's perspective of the energy consumption. However, the fact that renewable energy is increasing does not mean that conventional power like crude oil is decreasing. Quite the contrary process is the case. The usage and world oil demand is increasing constantly.

The industrial consumption rises and the consumption of renewable energy is increasing constantly. While solar, wind onshore, wind offshore, hydropower, and biomass are becoming more and more attractive to consumers, conventional energy like lignite is remaining the same. So there is an increasing demand in renewable energy over the years, yet the demand of conventional energy did not decrease (Fig. 2.1).

What has to be observed from this, is that the energy landscape is shaped by rising renewable energy sources and conventional resources are increasing too.

Oil is a limited and a finite resource which will be exhausted one day. Logically the worldwide demand should increase and the worldwide supply should decrease. As a result prices should also increase, because of the economic demand and supply principle.

Oil is needed for heating, transportation, electricity and is a prime raw material for chemicals, fertilizer, and plastics. This causes the demand continuing to increase constantly every year, which is one reason for oil companies to operate in this industry.

It is proved that oil will be depleted one day, but the current oil capacity in the world is still very high. There are still enough oil sources which will last for many decades. There will be an oil peak one day, but it is quite uncertain when this will happen. Additionally, the production of oil is developing and becomes more efficient due to an increasing level of innovation.

Oil production and supply has developed in the past 10 years. New and more efficient ways of producing oil has been implemented. Oil companies can, therefore, produce twice the amount of oil as before at the same time. Improvement of technology is driving the supply due to increasing efficiency and therefore increasing profitability for oil companies. Also the election of Americas president Trump supports the energy industry. President Trump is a pro energy and pro infrastructure oriented and therefore changed the game in the energy sector. This change helps the American energy sector, to reach their goal of energy independence. This power is also driving technology and innovation within the oil industry and contributes to efficiency and competitiveness of oil companies in the United States.

The amount of reserves could, therefore, be increased from 2.9 to 4.8 trillion barrels using the new invented technologies already available today. The calculated global demand by 2050 is 2.5 trillion barrels.

2.6 trillion barrels of oil are stored beneath the earth surface. That is enough to quench the world's oil demand twice by 2050. Basically, there is enough oil for everyone. However, the International Energy Agency (IEA) warns of an impending

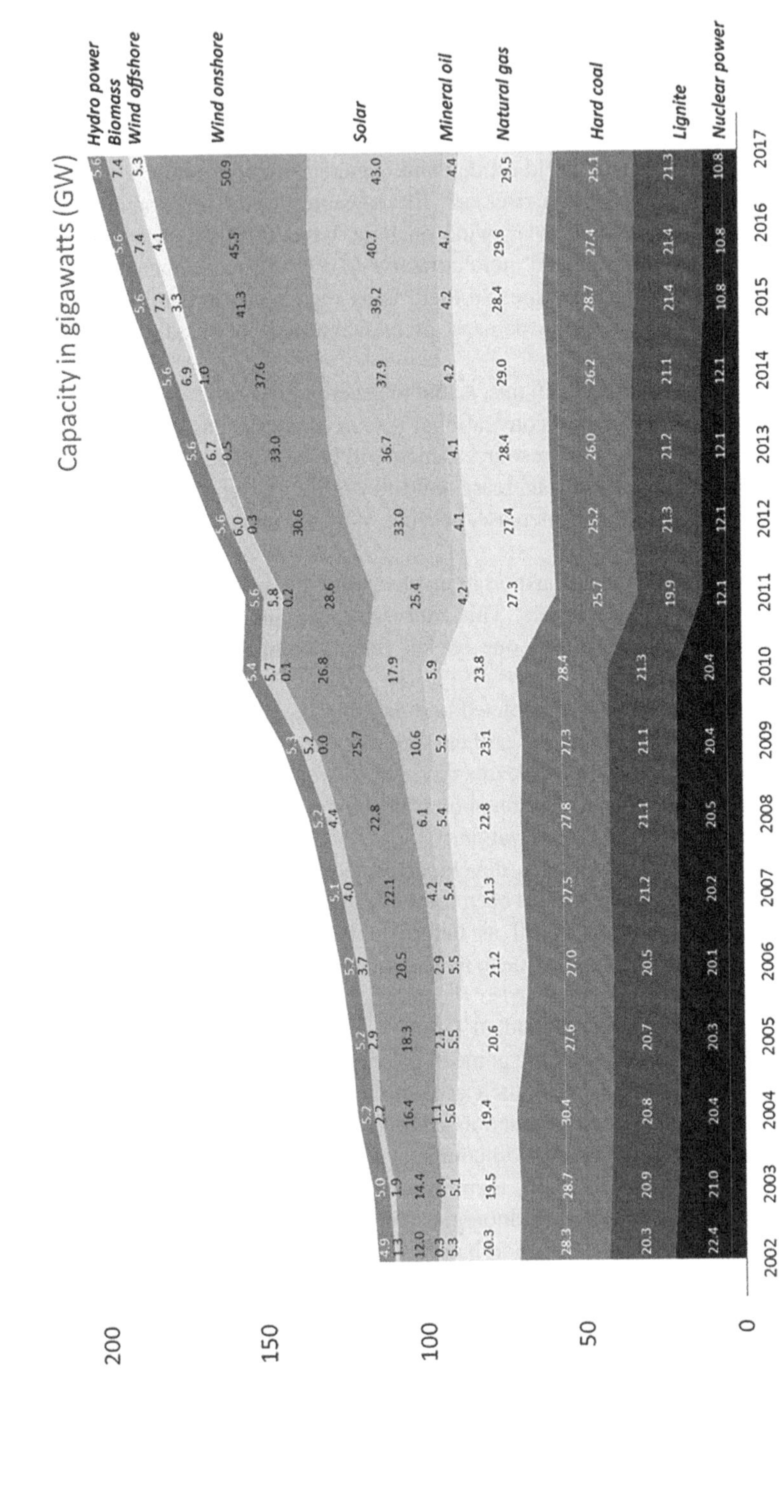

Fig. 2.1 Energy Generation Germany. Source: Appunn et al. (2018)

supply shortage in its report "Oil 2017" over the next five years (International Energy Agency, 2017).

Theoretically, the reserves are enough to supply the world. Whether that happens depends ultimately on the oil companies, because they decide how much they invest in new projects and technologies. In the past two years, the investment in research and development of oil companies has fallen drastically from 800 to 450 billion Euros per year. This means that the future oil capacity is strongly dependent on the companies itself (Streit, 2017).

An additional effect which is considerable in order to understand the demand and supply is the so-called Elon Musk effect. Every oil producing country is competing against milestones like Tesla. The invention of cars which run by electricity, is threatening the oil industry tremendously. The fact that new sustainable powers could replace oil is causing a big reaction from the oil industry. Innovations like e-cars are expensive to produce and therefore also expensive for Customer.

The oil industry strives for low prices in order to be more favorable than renewable energy. Low oil prices are setting high market entry barriers toward new innovational "non-oil" products. So it is not in the long-term interest of oil companies to increase oil prices, because it would drive gasoline prices back and this would lower the entry barriers for Elon Musk's company "Tesla" which is threatening the oil industry. Oil companies want to hold the oil prices low in order to get customers attracted and addicted to the oil. Otherwise the oil industry would not have a chance in the long-term view. The development of technology within the oil industry and the "Elon Musk effect" are two major ongoing trends, which just started yet and will persist in the next 10 years. The Elon Musk effect is causing a humongous price competition between renewable energies and the oil industry (Fox Business, 2016). These insights show why and how the oversupply is arising and causing a decreasing worldwide future oil prices in 2013–2015 (Vehlewald, 2015).

But not only the supply is increasing, also the demand. By looking at the world populations, one can observe a constantly increasing population. The next fact to obtain is the urbanization. The shift from living in rural areas to living in urban cities is increasing as well (Siegfried et al., 2021). More and more people are moving to urban cities and this is causing a rapid increase of urban population in the world. The urban population has significantly higher oil consumption than population in rural areas because cities do not have an efficient infrastructure yet. Concepts like car sharing and EVs are not well established, which makes urbanization an increasing force for conventional energies (Fig. 2.2).

In the following illustration one can see the predicted energy demand in urban cities and rural areas.

As a conclusion in between it can be said that worldwide demand and supply will continue to increase within the next 10 years due to the development of technology in the oil industry, Elon Musk effect, and urbanization and increasing population.

It has been identified that oil is a finite resource but will not run out in foreseeable time. Additionally, oil prices will not increase too strongly due to the competition with the renewable energy sector. The oil demand will increase too, due to

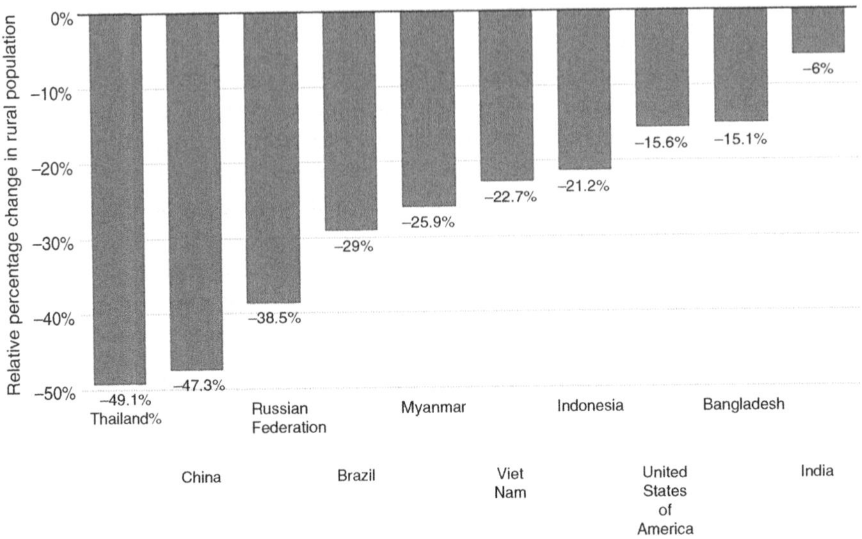

Fig. 2.2 Urbanization. Source: United Nations (2017)

population increases and urbanization. So the question rises why should the world than switch to EVs.

The Stone Age did not end, because the world run out of stones, it ended because of ideas, new technologies and innovations (Sears, 2010).

This quote helps to understand that it is not necessary to drive EVs because oil resources will run out. It is more because EVs are changing the old concepts of mobility. The oil era will not end because there is no oil anymore, it will more likely end because people will not need it or buy it anymore.

Car manufacturers are investing billions in the development of EV. It will have a longer range and will be faster, cheaper, safer, more convenient and will have a better infrastructure. This will lead to an increased customer preference for buying EVs. As a result more EVs will be on the road and the oil demand will decrease. There are two ways of causing an oil crash. One way is when oil companies are producing more oil than demanded which leads to an oversupply. The result is a very low oil price which makes it hard for the oil industry to generate earnings, which happened in 2013–2015. The second way to cause an oil crash is when the demand side is reducing its demand due to the usage of EVs. As a consequence the oil price will shrink again. This will happen if the EV sales will increase. Today EV market share is only 1%. However, EV companies are planning to increase EV share tremendously. Tesla sold approx. 50,000 EVs in 2016 and build a factory to produce and sell 500,000 EVs in 2020.

Assuming that Tesla and also other EV manufacturers are hitting their forecasts for EV, worldwide oil demand will shrink significantly and will cause an oil crisis. This could mean the end of the oil era at least for the car industry (Tesla, 2017).

This section should show that oil is not the crucial influencing factor for EVs. Moreover, it should show that the development of oil is strongly dependent on the future development of EVs. The future oil capacity is unpredictable and is likely to last many decades if oil companies are investing in new technologies. There are two basic oil scenarios (Randall, 2016).

Oil capacity will last long enough and EV market share will increase. The result is an oil crisis due too low oil demand.

Oil capacity will shrink and will be depleted. The result is a very high oil price which would attract people to drive EVs in order to overcome high oil prices (McKinsey & Company, 2016).

2.5 Customer Preferences

In order to analyze the progress of e-mobility, it has to be identified which factors are impacting customer preferences the most. These factors then will be evaluated in order to make a statement about the impact of customer preferences on the future development of e-mobility.

Therefore, the purpose of this chapter is to get an understanding of customer preferences and of which factors it consists of.

The awareness for climate change, technology leaps, and policy makers are all contributing to the successful development of today's EVs. However, customer preferences are an additional influencing factor of e-mobility, which has a humongous effect on the future progress of EVs. Without having customers who are willing to buy EVs, there will not be any car manufacturers who will longer build EVs. One argument which implies that OEMs will produce and further develop EVs is the influencing power of policy makers. Policy makers are forcing OEMs to produce EVs, but if OEMs cannot sell them due to missing demand, large car manufacturers will head into financial crisis with the consequence to go bankrupt. Due to the fact that there is no other well-developed alternative of ICEs and EVs, except of public transportation, many OEMs and suppliers would become insolvency because of high production costs and low revenue. This would lead to a cross-industry global crisis.

This scenario is of course a very simplified dramatical representation of what could happen if customer preferences will not imply any buying intentions toward EVs in the future. This example should make clear how important it is for OEMs and other stake-holders to focus very heavily on customer preferences and how to increase customers willingness to buy EVs.

In order to analyze the future progress of e-mobility, it is crucial to understand how the demand side impacts e-mobility. The demand side of e-mobility consists of car drivers and their customer preferences, which depend on numerous variables. There is a lot of literature and empirical models which analyze the customer preferences toward EVs. Each study and model first identifies numerous variables. Subsequently, these models analyze how these variables are correlating with the willingness to buy an EV. The purpose of these empirical models is to find out, which variables are impacting customer preferences significantly (Siegfried, 2013). All major studies are categorizing the variables almost similarly. Taking the most relevant variables into consideration, four different variable categories can be formed. They can be divided into financial, technical, policy, and individual variables (Liao et al., 2016).

The study from P3 group (2013) gives extensive information about the acceptance of EVs, which will be among others used for the analysis of the variables. Additionally, the meta-analysis of Liao et al. (2016) will give supported information about other relevant studies which have been executed in the field of customer acceptance.

2.5.1 *Financial Variables*

Financial variables refer to all financial aspects of the purchase and use of an EV. This category includes the economic advantage for customers. The acquisition costs but also maintenance and other expenses like fuel costs are very important influencing factors for the buy decision of customers. An EV has higher production costs and therefore a higher purchase price for customers. However, they have significantly lower maintenance and fuel costs and the high acquisition cost of an EV can be compensated with special financing solutions like leasing. Additionally, EVs can be a move to protect oneself toward the rising oil and fuel prices in the future. So there are several financial factors which contribute to the buy decision of customers. Nevertheless, the high price is a crucial counterargument for customers.

Most studies found out that the financial advantage has the strongest correlation with the buy decision of customers (Seipp et al., 2020). The stronger the customers are convinced that they can save money with EVs or be able to overcome rising fuel prices, the more they are willing to buy an EV. The comparatively high acquisition costs of EVs, on the one hand, but rising gasoline prices and low running costs on the other, makes it reasonable from a financial point of view to buy an EV.

One study also found out that there is a significant positive effect on the purchase and leasing preference. The research model used in the study is based on the theory of reasoned action, which is considered one of the most important theories for explaining human behavior. This theory says that attitudes toward behavior and subjective norms are crucial determinants of behavioral intentions and that actual behavior is significantly influenced by these intentions (Ajzen, 1985). The study reveals that people are more likely to purchase an EV, if there is the possibility of leasing an EV. This indicates that the financing solutions which are offered by OEMs or financial institutions have the potential for increasing EV acceptance.

The research model of P3 group clearly shows that both economic and environmental considerations play a major role in the purchasing decision of EVs. In particular, the still high purchase costs have the potential to decrease and can in return increase acceptance, since only about 24% of the persons surveyed are willing to pay a surcharge for EVs. About three quarters of surveyed people would only accept a surcharge of up to 3000 euros. However, considering the current additional costs in this area, it becomes clear that there is still a strong need for improvement. Tesla, Nissan, and Chevrolet are planning to sell EV in the next few years for around 25,000 euros, which would be equal to ICE cars (Randall, 2016).

In addition, when developing new business models, it should be kept in mind that there are quite different preferences with regard to the form of financing. 60% of respondents say they generally prefer to buy an EV, while about 18% would choose to lease an EV (Paternoga et al., 2013).

2.5.2 Technical Variables

The technical variables include the technical characteristics and also the user-friendliness as well as the personal mobility and charging requirements for EVs.

The range has a strong effect on the buying decision of customers. The battery capacity is one of the toughest challenges for OEMs to increase and it is one of the most important factors for customers. The most empirical research models which analyzed technical factors found out that "range" has statistically a positive effect on customers buying decision.

BearingPoint and PP: Agenda, one of the leading communication agencies for e-mobility, surveyed 500 persons who plan to buy a new car within the next 12 months. According to their findings, one of the biggest barriers is the low range of EV. 69% of surveyed participants say that the range is a reason for not buying EVs. 73% answered that the charging infrastructure is another reason which makes them buying an ICE instead of an EV (Bock, 2017).

McKinsey surveyed 7000 consumers in the United States, Norway, China, and Germany. McKinsey's research result shows that around 25% of all participants do not buy a car because of the range and another 20% is not willing to buy an EV due to the charging infrastructure. McKinsey research results strengthen the findings of the previous empirical models mentioned before (Gropp, 2017).

However, another study of the institution for center automotive research in cooperation with the university Duisburg-Essen (2012) has shown that range is not the crucial factor for customers.

Over a period of three months the behavior toward EV was analyzed in an experiment with 226 representatively selected persons. The willingness to buy an EV was low, yet the price was the biggest obstacle. The infrastructure of the loading points and the loading times are considered as additional weaknesses for the participants. Participants think that the range of EVs is sufficient and does not impact their buying decision. Due to the fact that the range is strongly depended on the factors like place of residence, lifestyle, and driving habits, studies which examine the range of EVs are difficult to compare and do not provide sufficient validity and comparability.

But the experiment of the central automotive research shows that the test group reacted very euphorically to EVs after test drives. After driving an EV and receiving instructions for the charging process, 71% of all test persons stated to consider EV when they buy their next cars. This is a very high value, demonstrating the importance of product testing in the implementation phase of EVs. 23% of respondents said they would consider EVs when making a purchase decision in the next four years (Dudenhöfer et al., 2012).

The research of P3 group also found out that persons who have already been able to gain experience with an EV during a test drive have higher buying intentions. From this it can be assumed that people who are about to make a car purchase decision are more willing to buy an EV when they have gained further information and experience with EVs.

This finding underlines the high importance of concepts like car sharing, which increase the familiarity between EVs and potential new customers.

With regard to the direct use of an EV, the research of P3 group found out that the perceived user-friendliness has a positive influence on the purchase intention. As a result, the willingness to purchase an EV increases when potential customers are convinced that no additional technical knowledge is required or that charging is not dangerous.

The personal mobility and charging requirements in the research have the only significant negative correlation with the willingness to buy. It shows that persons with higher expectations toward the charging process and their personal mobility in terms of reach and range have a lower purchase intention (Liao et al., 2016; Paternoga et al., 2013).

2.5.3 Regulation Variables

Policies and regulations also have an impact on the buying decisions of customers. There are several policies which are contributing customers to buy EVs.

Policy makers are paying buyers' premium and reduced the purchase tax for customers who are buying EVs in order to reduce the overall purchase price. Additionally, EV drivers benefit from non-financial polices like using priority, bus, and express lanes to overcome traffic and crowded roads (Handelsblatt, 2015).

In some counties like the UK, EVs drivers also benefit from free parking rights (BBC, 2016).

According to a German study called "Autotrends," which surveyed more than 1000 people, around 35% of surveyed participants would be motivated to buy an EV, because of the buyer's premium (Credit Plus Bank AG, 2016).

In contrast, other studies from the meta-analysis and P3 Group have shown that non-financial policies have no significant effect on the willingness to buy EVs. The buyers' premium and the tax reduction have a significant effect on the buyers' decision. However, these studies again are not providing sufficient reliability and validity due to the different circumstances of the participants. A participant who lives in a large urban metropole would validate the access on express lanes and free parking right higher than someone living in a city where traffic is moderate. Therefore, individual variables play an important role for the buyer's decisions.

2.5.4 *Individual Variables*

Individual variables are also impacting buying decisions significantly.

In addition to the external influence factor, the influence of personal attitudes could be proven in the study of P3 group. For example, as expected, the environmental awareness of customers positively influences their willingness to buy EVs. As a result, the potentially positive environmental impact of e-mobility with regard to the protection of fossil fuels and the reduction of greenhouse gas emissions is also perceived by customers. In addition, there is a positive correlation between the willingness to innovate and the intention to buy. This result suggests that respondents perceive EV primarily as an innovative overall system and less as a technical innovation (Paternoga et al., 2013).

The meta-analysis of Fanchao Liao, Eric Molin, and Bert Van We (2016) compares numerous studies toward individual factors regarding the customer preferences for EV. The meta-analysis shows the relevant individual based factors which are influencing the intention to buy a EV. For every individual factor, the research is looking at several relevant studies and evaluates the impact of each individual based factor on the willingness to buy an EV. The following factors have been identified by the meta-analysis: socio-demographic factors, psychological factors, and social influence.

Socio-demographic factors include gender, age, income, education level, and household. This factor is so far uncertain whether their effects are positive or negative. Studies could not find any correlation between this factor and customer preferences. All considered studies provide different outcomes. Thus, no uniform statement can be made for this factor.

Psychological factors are related to emotional behavior which influences customer preferences. One emotional factor has already been mentioned. Customers are more likely to buy an EV if they had a test drive and a further experience with driving an EV. The meta-analysis also found out that test drives are causing a psychological effect on the driver which leads to emotional thinking. The driver does not decide based on rational thinking but on emotional behavior. The meta-analysis shows a positive correlation between driving experience and customer preferences.

An additional psychological factor is the environmental motivation of consumers. The environmental friendly belief of consumers has a positive correlation with customer preferences. The awareness of climate change is motivating consumers to drive innovative alternative powertrains like EVs.

Another emotional factor which has been identified is the symbol and status of EVs. It is still difficult to assess the emotional bond between customers and EVs. The question arises, if EVs are decreasing the image of a car as a status symbol, which is hard to answer because of complex cultural aspects which have to be considered (Cornet et al., 2012). From recent news customers in the USA are connecting EVs with financial wealth and social prestige (Eisenstein, 2013).

Additionally, several studies have found out that social influence has a significant correlations with customer preferences. Personal network and social interactions between friends, colleagues, and family members are impacting the buy intention of an individual significantly (Liao et al., 2016).

References

Ajzen, I. (1985). From intentions to actions: A theory of planned behavior. In J. Kuhl & J. Beckmann (Eds.), *Action control* (pp. 11–29). Springer.

Appunn, K., Bieler, F., & Wettengel, J. (2018, April 3). *Germany's energy consumption and power mix in charts*. Retrieved September 20, 2021, from Clean Energy Wire: https://www.cleanenergywire.org/factsheets/germanys-energy-consumption-and-power-mix-charts

BBC. (2016, November 28). *Free city parking for pure electric vehicles*. Retrieved September 20, 2021, from BBC News: http://www.bbc.com/news/uk-scotland-tayside-central-38130094

Bitkom e.V. (2015, December 11). *Energieeffizienz in Rechenzentren*. (T. u. Bundesverbanbozemd Informationswirtschaft, Ed.) Retrieved September 20, 2021, from Bitkom Website: https://www.bitkom.org/noindex/Publikationen/2015/Leitfaden/LF-Energieeffizienz-in-Rechenzentren/150911-LF-Energieeffizienz-in-RZ.pdf

Blume, J. (2015, November 25). *Verkehr verursacht fast ein Viertel der weltweiten CO2-Emissionen*. Retrieved September 20, 2021, from Heise Website: https://www.heise.de/tp/features/Verkehr-verursacht-fast-ein-Viertel-der-weltweiten-CO2-Emissionen-3376825.html?seite=all

Bock, A. (2017, July 5). *Bearing Point: Pressemitteilung: Nächstes Auto. E-Auto*! Retrieved May 2, 2018, from Bearing Point Website: https://www.bearingpoint.com/de-de/ueber-uns/pressemitteilungen-und-medienberichte/pressemitteilungen/e-auto-news/

California Air Resources Board. (2012). *Climate change programs*. Retrieved September 20, 2021, from California Air Resources Board Website: https://www.arb.ca.gov/cc/ab32/ab32.htm

Conrady, R., & Buck, M. (2012). Status quo and future prospects of sustainable mobility. In *Trends and issues in global tourism 2012* (pp. 240–249). Springer.

Cornet, A., Mohr, D., Weig, F., Zerlin, B., & Hein, A. (2012, February). *Mobility of the future*. Retrieved September 20, 2021, from McKinsey & Company: https://www.mckinsey.com/~/media/mckinsey/dotcom/client_service/automotive%20and%20assembly/pdfs/mobility_of_the_future_brochure.ashx

Credit Plus Bank AG. (2016). *Bevölkerungsbefragung: Autotrends 2016*. Retrieved September 20, 2021, from Credit Plus Bank AG: https://www.creditplus.de/fileadmin/03_Ueber_Creditplus/Newsroom_und_Pressebereich/Newsroom/CP_160427_Kurzbefragung_Automobilitaet_2016.pdf

D'Aprile, P., Newman, J., & Pinner, D. (2016, August). *The new economics of energy storage*. Retrieved September 20, 2021, from McKinsey & Company Website: https://www.mckinsey.com/business-functions/sustainability-and-resource-productivity/our-insights/the-new-economics-of-energy-storage

Desmond, K. (2016). *Innovators in battery technology*. McFarland & Company.

Deutsche Bank. (2011, October). *Electromobility*. Retrieved September 20, 2021, from Deutsche Bank Research: http://www.dbresearch.com/PROD/DBR_INTERNET_EN-PROD/PROD0000000000279687/Electromobility%3A+Falling+costs+are+a+must.pdf

Dudenhöfer, F., Dudenhöffer, K., & Bussmann, L. (2012). *Elektromobilität braucht intelligente Förderung*. Retrieved September 20, 2021, from Universität Dusiburg-Essen: https://www.uni-due.de/~hk0378/publikationen/2012/Wirtschaftsdienst_4-2012_S274-279_kl.pdf

Eisenstein, P. (2013, November 4). *Tesla electric car becoming a must-buy for the wealthy*. Retrieved September 20, 2021, from NBC News: https://www.nbcnews.com/businessmain/tesla-electric-car-becoming-must-buy-wealthy-8C11522443

Ernst & Young Global Automotive Center. (2012). *Beyond the plug: finding value in the emerging electric vehicle charging ecosystem*. Retrieved September 20, 2021, from Ernst & Young Website:

European Commission. (2017). *Reducing CO2 emissions from passenger cars*. Retrieved September 20, 2021, from European Commission Website: https://ec.europa.eu/clima/policies/transport/vehicles/cars_en

Fox Business. (2016, November 28). *The Elon Musk effect on oil demand*.

Funda, P. (2014, August). *Globales Wachstum Chance oder Risiko*. Retrieved September 20, 2021, from pwc Website: https://www.pwc.de/de/automobilindustrie/assets/pwc_studie_automotive_globales-wachstum.pdf

Giesen, C., Peking, & Hägler, M. (2017, September 18). *China führt Quote für E-Autos ein*. Retrieved September 20, 2021, from Süddeutsche Zeitung: http://www.sueddeutsche.de/wirtschaft/e-mobilitaet-china-fuehrt-quote-fuer-e-autos-ein-1.3687137

Greenpeace. (2014). *Folgen des Klimawandels*. Retrieved September 20, 2021, from Greenpeace Website: https://www.greenpeace.de/themen/klimawandel/folgen-des-klimawandels

Gropp, M. (2017, January 5). *Wieso die Deutschen mit Elektroautos hadern*. Retrieved September 20, 2021, from Frankfurter Allgemeine Website: http://www.faz.net/aktuell/wirtschaft/neue-mobilitaet/warum-deutsche-gegenueber-elektroautos-skeptisch-sind-14603445.html

Handelsblatt. (2015, September 26). *Neue Privilegien für Elektroautos*. Retrieved September 20, 2021, from Handelsblatt Website: http://www.handelsblatt.com/auto/ratgeber-service/neue-privilegien-fuer-elektroautos-freie-fahrt-fuer-tesla-und-co/12372620.html

Hotten, R. (2015, December 10). *Volkswagen: The scandal explained*. Retrieved September 20, 2021, from BBC News: https://www.bbc.com/news/business-34324772

Intergovernmental Panel on Climate Change. (2014, December 1). *Climate Change 2014: Synbook Report*. Retrieved September 20, 2021, from Intergovernmental Panel on Climate Change: http://www.ipcc.ch/report/ar5/syr/

International Energy Agency. (2017). *Oil 2017*. Retrieved September 20, 2021, from International Energy Agency: https://www.iea.org/Textbase/npsum/oil2017MRSsum.pdf

International Organization of Motor Vehicle Manufacturers. (2018). *2017 Statistics*. Retrieved September 20, 2021, from International Organization of Motor Vehicle Manufacturers: http://www.oica.net/category/production-statistics/2017-statistics/

Lempp, M., & Siegfried, P. (2021). *Automotive disruption and the urban mobility revolution – Rethinking the business model 2030*. Business guides on the go. Springer. ISBN: 978-3-030-90035-9. https://doi.org/10.1007/978-3-030-90036-6

Liao, F., Molin, E., & Van We, B. (2016). Consumer preferences for electric vehicles. (I. UK, Ed.). *Transport Reviews, 3*(37), 252–275.

Maxwill, P. (2012, July 11). *Elektroauto-Revolution* 1912. Retrieved September 20, 2021, from Spiegel Online: http://www.spiegel.de/einestages/elektroauto-revolution-vor-100-jahren-a-947600.html

McKinsey & Company. (2016, February). *Beyond the supercyle*. Retrieved September 20, 2021, from McKinsey & Company Website: https://www.mckinsey.de/files/170216_pm_mgi_beyond_the_supercycle.pdf

Mokyr, J., & Strotz, R. (1998, August). *The second industrial revolution, 1870–1914*. Evanston.

Paternoga, S., Pieper, N., Woisetschläger, D., Beuscher, G., & Wachalski, T. (2013). *P3 Group - Akzeptanz von Elektrofahrzeugen*. Retrieved September 20, 2021, from Technische Universität Braunschweig: https://www.tu-braunschweig.de/Medien-DB/aip-ad/veroeffentlichungen/elektromobilitatesstudie.pdf

Randall, T. (2016, February 25). *Here's how electric cars will cause the next oil crisis*. Retrieved September 20, 2021, from Bloomberg website: https://www.bloomberg.com/features/2016-ev-oil-crisis/

Rohrer, P. (2017). Technik. In R. A. Revue (Ed.), *Automobil Revue* (pp. 14–22). Schwabe AG.

Roser, M. (2018). *Future population growth*. Retrieved September 20, 2021, from https://ourworldindata.org/future-population-growth

Sears, R. (2010). *Planen für die Zeit nach dem Öl*.

Seipp, V., Michel, A., & Siegfried, P. (2020). Review of international supply chain risk within banking regulations in Asia, US and EU including proposals to improve cost efficiency by meeting regulatory compliance, journal financial risk management (JFRM). *C-Journal*. https://doi.org/10.4236/jfrm.2020.93013

Siegfried, P. (2013). *The importance of the service sector for the industry*. Teaching Crossroads: 9th IPB Erasmus Week, Instituto Politécnico de Braganca. ISBN: 978-972-745-166-1 (S. 13–23).

Siegfried, P., Michel, A., Tänzler, J., & Zhang, J. (2021). Analysing sustainability issues in urban logistics in the context of growth of e-commerce. *Journal of Social Sciences, IV*(1), 6–11. ISSN: 2587-3490.

Siegfried, P., & Strak, D. (2021). Grüne Logistik: Eine Untersuchung ausgewählter al-ternativer Antriebstechnologien im Güterverkehr, Zeitschrift für Verkehrswissen-schaft (ZfV). D-Journal. ISSN: 0044-3670.

Stapleton, A. (2013, August 8). *Technologiewechsel im Akku-Bereich*. Retrieved September 20, 2021, from Smarter World Website: http://www.smarterworld.de/smart-power/batterien/artikel/100504/).

Streit, M. (2017, July 6). *Energieagentur IEA warnt vor Engpass*. Retrieved September 20, 2021, from Handelsblatt website: http://www.handelsblatt.com/finanzen/maerkte/devisen-rohstoffe/oelmarkt-energieagentur-iea-warnt-vor-engpass/19479206.html

Tesla. (2017). *Tesla Werk*. Retrieved September 20, 2021, from Tesla Website: https://www.tesla.com/de_DE/factory?redirect=no

Teslarati Network. (2013, July 19). *Tesla model S weight distribution*. Retrieved September 20, 2021, from Teslarati website: https://www.teslarati.com/tesla-model-s-weight

T-Online. (2017, November 21). *Wie EU-Staaten den Kauf von E-Autos ankurbeln*. Retrieved September 20, 2021, from T-Online Website: https://www.t-online.de/auto/elektromobilitaet/id_82745454/elektroautos-wie-eu-staaten-den-kauf-von-e-autos-foerdern.html

United Nations. (2017). *World population prospects*. Retrieved September 20, 2021, from Urbanization: https://esa.un.org/unpd/wpp/DataSources/

Vaughan, A. (2015, November 9). *Earth's climate entering new 'permanent reality' as CO2 hits new high*. Retrieved September 20, 2021, from the guardian: https://www.theguardian.com/environment/2015/nov/09/earths-climate-entering-new-permanent-reality-as-co2-hits-new-high

Vehlewald, H. (2015, December 9). *Warum wird das Öl immer billiger?* Retrieved September 20, 2021, from Bild Wirtschaft: https://www.bild.de/geld/wirtschaft/wirtschaft/warum-wird-oel-immer-billiger-43732938.bild.html

Verband der Automobilindustrie. (2015). *CO_2-Regulierung bei Pkw und leichten Nutzfahrzeugen*. Retrieved September 20, 2021, from Verband der Automobilindustrie Website: https://www.vda.de/de/themen/umwelt-und-klima/co2-regulierung-bei-pkw-und-leichten-nfz/co2-regulierung-bei-pkw-und-leichten-nutzfahrzeugen

Verband der Automobilindustrie. (2017, December). *CO_2-Regulierung bei Pkw und leichten Nutzfahrzeugen*. Retrieved September 20, 2021, from VDA Website: https://www.vda.de/de/themen/umwelt-und-klima/co2-regulierung-bei-pkw-und-leichten-nfz/co2-regulierung-bei-pkw-und-leichten-nutzfahrzeugen.html

Wilkens, A. (2017, August 11). *SPD-Kanzlerkandidat Schulz fordert verbindliche quote für Elektroautos in Europa*. Retrieved September 20, 2021, from Heise website: https://www.heise.de/newsticker/meldung/SPD-Kanzlerkandidat-Schulz-fordert-verbindliche-Quote-fuer-Elektroautos-in-Europa-3797709.html

Chapter 3
Development of E-Mobility

3.1 Global E-Mobility Market

In the previous chapter, relevant influencing factors have been identified and analyzed already. It showed why and how the identified drivers are impacting e-mobility. This chapter will include the pre identified influencing factors and will link them to the future development of e-mobility. From the previous chapter, it can be seen that customer preference is one key driver for e-mobility. Additionally, political regulations, technological process (development of battery), and the charging infrastructure will play an important role in order to forecast the e-mobility progress.

According to a study of the Center of Automotive Management global EV sales have increased significantly in every EV core market (see Appendix A) (Center of Automotive Management, 2017).

China increased the EV and PHEV sales in 2017 to more than 777,000, including commercial vehicles and buses. The market share of EVs rises to 2.7% of new registrations. In particular, the number of purely EVs (BEVs) has increased and remains with 652,000 units (84%) by far higher than the plug-in hybrid (PHEV), with 125,000 units (16%).

In the USA, EVs recorded a solid growth of 24% in 2017. The market leader is Tesla: Almost every second EV is a Model S or Model X (currently the first Model 3 is added). Only second is General Motors with the Bolt model, followed by the Nissan Leaf.

Norway remains the European leader of EV sales: Norway has more than 62,300 EVs (+39%) after the full year, bringing the market share of EV to 39.3% of new registrations, with financial benefits such as the elimination of VAT, import tax, and vehicle tax; electric vehicles in Norway are often cheaper than the counterpart with internal combustion engine.

In Germany, 54,492 EVs (+117%) were sold, doubling their market share from 0.8 to 1.6%. Sales of plug-in hybrids are significantly higher than EV sales. Plug-in

K. Y. Ciftci et al., *The Potential Impact of E-Mobility on the Automotive Value Chain*, SpringerBriefs in Business, https://doi.org/10.1007/978-3-030-95599-1_3

hybrids sales increased to 29,436 (+114%), while the pure EV sales rise to 25,056 cars (+120%) (Viehmann, 2018).

Although ICEs make up 96% of European vehicle sales currently, it can be observed that EV growth rates are increasing consistently in every key market (China, the USA, Europe) (The Boston Consulting Group, 2017).

A research report of the Bank of America Merrill Lynch shows the increasing market share of EVs since 2015.

EVs and MHVs (HEV and PHEV) make up 3.6% of total vehicle sales in 2017 while ICEs are decreasing its market share but still hold the major market share of approx. 95.7% vehicles. It can be observed that the sales of EVs and MHVs overperformed the overall worldwide vehicle sales which grew 2.4% in 2017, while EVs and MHVs grew 25.7%.

Breaking sales figures down to purely EVs (BEV) without PHEVs and HEVs, it can be seen that EV sales increased from 350,000 in 2015 to 490,000 in 2016 with China holding the biggest share again of 51%.

China outpaced the US EV market in 2015 and is the biggest EV markets. China sold 204,000 NEVs (PHEVs and BEVs) and the USA 114,000 NEVs in 2015.

From review it can be observed that 76% of the total NEV sales from China come from purely EV sales (BEV), the rest of 24% come from PHEV sales. So China is highly focusing on the BEV market currently. European markets like Germany, the UK, and the Netherlands are depending on PHEV sales rather than BEV sales (Morgan Stanley, 2017, p. 32).

From the given market representations above, the following three main findings can be concluded:

- The three core markets of EVs are China, the USA, and Europe.
- EVs (EV, PHEV, HEV) are constantly increasing market share and represent a global growth market.
- China is leading the global EV market with a strong focus on BEVs rather than on PHEVs.

In order to get a basic understanding of the key OEMs, the following table shows the EV sales classified by OEMs and regions which are ranked by the sales volume of EVs.

Tesla leads by far the EV sales of American OEMs and holds approx. 89% of total EV sales from American manufacturers. In Europe, Renault and VW group are predominantly European OEM sales leaders followed by BMW with 13,267 sold EVs in 2016. In the Asian market, Nissan with 47,523 EVs and BYD with 43,309 EVs are dominating the EV sales of Asian automobile manufacturers in 2016.

3.2 Report Review

In order to create a forecast model, relevant drivers have to be analyzed, which will then help to predict the development of EVs. The most important drivers of e-mobility have already been identified in the previous chapter. This chapter will pick out the most relevant indicators of each driver and provide a deeper insight in those drivers. Therefore, exclusive reports from Morgan Stanley, Bank of America Merrill Lynch, J.P. Morgan, and Deloitte will be outlined and evaluated. All models can be found within the Appendix. These models then will be summarized in order to create an own forecast model.

Multiple forecast models and other indicators for the future development of e-mobility derive from reports and studies of research teams of renowned consultancies, institutions, and banks.

Each model takes different indicators into account and has partly different perspectives on the e-mobility development. However, there are some basic factors which can be found in almost every report. Regulatory changes, customer preferences, cost of battery, cost of ownership Infrastructure/charging station development, commodity implications, and mobility services have been analyzed in every model in one or the other way. This shows the high importance of these factors regarding the development of electric vehicles.

3.2.1 Forecast Model: Morgan Stanley

According to the study from Morgan Stanley, in 2050, the global EV fleet will contain one billion vehicles.

Their model is considering three main drivers and three key assumptions for their forecast.

1. *Cost of batteries will decrease within the future*

   So far, the batteries are one of the biggest cost factors in the production of EV. It is one key driver for the price of an EV and, therefore, also a key driver for the customer acceptance of EVs. According to a study from McKinsey, currently battery costs are about 195 euros per kilowatt-hour (kWh). With a typical battery size of 60 kWh, this means that there will be an additional cost of about 11,700 euros compared to an ICE (McKinsey & Company, 2017).

   The prices of batteries for EVs have dropped by 80% in the last 6 years and will continue to fall. This will lower the market prices for EVs and make them more attractive for consumers. In order to accelerate this development, Tesla has built its own battery factory in Nevada together with Panasonic, where battery cells are manufactured in-house (Tesla, 2014).

   By this move, Tesla becomes independent of major battery manufacturers such as LG and Samsung. This gives Tesla the possibility to lower the world market prices for EV batteries also for the future.

Looking at Morgan Stanley's study, they expect that battery costs will decrease to 84 euros until early 2020. Their expectation is based on two arguments. The first argument points out that technology and further developments will make battery production more efficient. The second argument compares the battery prices with the development of solar panels which fell from 60 euros per watt to less than 1 euro within 40 years. Similar can be observed with the cost of lithium-ion batteries which has fallen from over 1000 euros per kWh to below 195 euros today after its initial set-up. Comparing the battery development to solar panels Morgan Stanley expects lithium-ion battery costs per kWh to easily fall below 84 euros by 2025.

Due to the expected rapid reduction of battery cost, car manufacturers can sell EVs for a competitive consumer price. This will lead to higher customer acceptance which increases EVs sales.

2. *Consumer preferences will shift toward EVs*

   Additional to battery costs, customer preferences are a further driver for Morgan Stanleys forecast model. As mentioned within the previous chapter, customer preferences have a significant impact on the customers buying decision for EVs. According to MS there are several developments which increase customers' acceptance for electric cars. These developments include the improving technology, which is extending the range of electric cars. Technology also enables faster charging times and a broader charging infrastructure. Governmental regulations come with many consumer benefits such as lower road taxes and are also contributing to customers' acceptance. Furthermore, electricity charging costs for EVs are lower than fuel costs like petrol and diesel.

   The general German automobile club "ADAC" published an extensive table with the costs of the EV and hybrid models currently available on the market. The table compares the cent price per kilometer with the fuel costs of equally diesel or petrol cars. From this table it can be observed that a higher distances traveled each year, the more likely electricity charging costs are lower than diesel or petrol costs (Allgemeiner Deutscher Automobil-Club e.V., 2018).

3. *Consumer price and finance cost will decrease in future*

   The third big driver for their forecast model is the falling consumer prices. Acquisition costs for EV nowadays are still higher than for ICEs. However, there are two factors impacting the current situation. The first factor is regulatory requirements which force OEMs to sell more hybrid and electric cars in order to meet fleet emission requirements. This will make OEMs compete with car prices against each other. As a result the consumer benefits from this price competition. There are already competitive EV manufacturers with very competitive prices in 2018. The Nissan Leaf costs approx. 30,000 euros and the Renault Zoe costs approx. 20,000 euros. In 2018, there will be more cars in this price range which will increase EV sales due to attractive prices (Schweitzer, 2018).

   Based on these key drivers Morgan Stanley forms the following three assumptions for their forecast model:

1. Global car sales are dependent on *population growth, replacement frequency* of car fleets, and *ownership changes.*
2. *Global Population* continues to increase which will increase global car ownership.
3. Due to urbanization shared mobility concepts will increase. Shared car fleets reduce the *ownership* of the global car fleet, but do not reduce the sales volume. Shared riding concepts increase the mileage driven of a car which leads to faster depreciation of car fleets. As a result the *replacement frequency* of cars will rise due to a higher usage of cars.

Based on the three key drivers and three key assumptions MS created their forecast model which presents the future global car sales until 2050. Within their base case they are expecting global electric car sales to be growing by 50% to over 130 million a year, and predict that EVs will make up to 80% of global sales by 2050. Main reason is the 50% population growth and higher ownership which both lead to 50% growth in electric car sales. This will lead in total to a doubling of the global vehicle fleet to over two billion cars.

The model MS predicts the world car ownership change until 2050 which is based on the GDP per head. World car ownership will increase from 15% to 25% ownership and will therefore also increase global car sales.

It can be observed that a higher GDP per head can be connected with higher car ownership.

Additionally, MS predicts that ICEs have the potential to peak below 100 million cars within the next 3–5 years. The year of intersection where world EV sales surpasses world ICE sales is predicted to be 2038 when global EV sales make up about 50% of the worlds sales of cars. The share of EVs of the total car fleet will be 57% by 2050.

3.2.2 *Forecast Model: UBS*

In the following course of this chapter, additional studies will be given in order to form an own forecast model. Due to limitations of this Book, the following presentations of the forecast models will be limited. The selected forecast model will be briefly described, shown in a visual graph. Information about a more detailed outline of the reports and models can be found within the appendix.

UBS forecast model analyzes the potential EV development until 2025.

UBS predicts make EVs take 15.8% of global new car sales in 2025. The biggest share of EVs can be attributed to purely BEVs, because the expected battery costs in 2025 will fall and as a result the BEV will be cheaper to manufacture than the PHEV. Assuming UBS's forecast model to be right, almost every sixth purchased car in the world will be an EV by 2025.

The following diagram shows the time in years on the x-axis and the sales of EVs in million units on the y-axis. This research will only take the base case into consideration.

UBS analyzed the main reasons for customers not to buy EVs. The main reasons which have been identified contain the battery life, purchase price (cost of ownership), and access to charging. For each reason UBS has analyzed the future outlook.

The battery life of EVs is very dependent on the car brand, however the average battery life is getting higher year by year. The battery capacity of the Tesla S model reaches approx. 450,000 km which are equivalent to over 1000 full charges. Tesla plans to even double this amount to 900,000 km.

The purchase price will decrease year by year and will intersect with the ICE purchase price by 2025. UBS predicts that the total cost of ownership of EVs will be cheaper than of ICEs even before 2025 due to lower energy and maintenance costs.

Access to charging is accelerating currently as many mass market EVs are to be launched within the next 3 years. The new EVs will have a range of approx. 400–480 km, therefore charging infrastructure investments are going to accelerate. Additionally, governments are supporting infrastructure investment. The German government has supported the charging infrastructure with 400 million euros in 2016 (Bundesministerium für Wirtschaft und Energie, 2014).

Concerning the outlook of these factors, it can be observed that the reasons of not buying an EV for customers will shrink rapidly and EV demand will overperform ICE demand by 2025 already.

UBS summary: Important facts and figures

- Sales in units by 2025: 16.2 million EVs
- Share of total sales by 2025: 15.8%
- Total sales by 2025: 103 million cars

3.2.3 Forecast Model: J.P. Morgan

Compared to UBS and Morgan Stanley forecast models, JP Morgan is forecasting weaker EV sales figures for 2025 and 2030. This is caused by the evaluation of infrastructure and cost of battery. JP Morgan estimates that the development of the charging infrastructure will happen in a much slower pace than in UBS forecast. Therefore, JP Morgan believes that PHEV will take a large share of the EV market as a bridge to purely EVs until infrastructure and battery costs have fallen significantly. This is estimated to happen by 2021. After 2021, PHEVs will almost reach their peak and purely EVs sales will start to accelerate rapidly.

Due to improving charging infrastructure and decreasing battery costs, demand for PHEV will reach its peak in 2028.

J.P. Morgan summary: Important facts and figures

- 2025: total EV market/sales: 3.9 million (4% of total demand)
- 2030: total EV market/sales: 9 million (8% of total demand)

- Important factors: regulatory government trends (e.g., fuel economy), manufacturing costs (mainly for batteries), building up recharging infrastructure

3.2.4 Forecast Model: Bank of America Merrill Lynch

The study "Global Electric Vehicle Primer: Fully charged by 2050" of Bank of America Merrill Lynch sharply analyzed the forecast of EVs based on identified supportive factors. Supportive factors are accelerating the EV development in their model. Bank of America Merrill Lynch supportive factors used in their study include the increased EV driving range, the decreasing EV manufacturing costs (2016–2020: $16,201 to $11,889) and increasing ICE manufacturing costs (EU/USA), supporting subsidies in the USA, ambitious EV sales targets for China, falling diesel sales/ residual vehicle pricing in EU and reduced EV charging infrastructure costs.

Bank of America Merrill Lynch summary: Important facts and figures

- 2025: 13,560,000 EVs sold (12% of total car sales)
- 2030: 40,460,000 EVs sold (34% of total car sales)
- 2050: 107,100,000 EVs sold (90% of total car sales)

3.2.5 Forecast Model: Deloitte

Deloitte forecast model is heavily focused on the customer preferences. Like the previous studies presented before, Deloitte is analyzing the customer preferences deeply and bring them together with determination parameters (influencing factors). In their study they are clustering customers into different customer groups. Deloitte differentiates six types of EV customers: (1) environmentally conscious "early adopters," (2) innovation and status-oriented "early adopters," (3) hesitant, (4) followers, (5) indifferent customers, and (6) "low-end consumers." These groups vary in size and enter the market on different timing. Currently, electric cars are still being bought by "early adopters" according to Deloitte's study "Mobility 2.0." Besides customer preferences, Deloitte analyzed two different key influencing factors for the development of e-mobility which is the cost of battery and the cost of ownership of ICEs and EVs (Deloitte 2017a, b).

So Deloitte is combining the findings of the analysis of customer preferences with the two influencing factors of e-mobility and created the following forecast model. The red bars represent ICEs and blue bars fuel cell cars.

Deloitte summary: Important facts and figures

- 2020: EV 3% of total Market share
- 2025: EV 17% (16.8) of total Market share
- 2030: EV 60% of total Market share

3.3 EV Market Share Forecast

Based on the presented studies and forecast model, the average market share of EVs has been calculated and has led to the following results (Tables 3.1 and 3.2).

Every model is based on different sales growth figures. Thus, the percentage only provides restricted information about the actual volume of cars sold. 4% EV market share in Morgan Stanley's forecast model provides a different EV sales volume in units than 4% in UBS model, for example. This is because the studies are based on different global car sales predictions. Therefore, an additional table will give an outlook of EV sales market share in million units. As a counterpart the figures have been also calculated for ICE sales in percent in Table 3.3. Finally, Table 3.4 sums up the average sales figures of ICEs and EVs in million vehicle units.

Based on these tables an overall forecast model has been developed. Figure 20 shows the ICE and EV sales share in percentage and Fig. 21 shows the applied data in million vehicle units (Fig. 3.1).

Based on the average sales figures from the previous analyzed reports, the following forecast model has been developed. It can be observed that EVs will surpass the global ICE sales by 2035 and will reach a total market share of 85% by 2050. The exact numeric values of the forecast models can be found in Fig. 3.2.

The figure above shows the data from the forecast model in million units. McKinsey executed a study, which worked out the potential revenue of the automobile industry. The McKinsey report showed that the automobile market will be a

Table 3.1 Market share EV sales

Release date	Model/Study	2020	2025	2030	2040	2050
September 2017	Morgan Stanley	4.0%	9.0%	16.0%	64.0%	80.0%
November 2017	UBS	–	15.8%	–	–	–
November 2017	JP Morgan	2.0%	3.8%	8.0%	–	–
March 2018	Bank of America Merrill Lynch	2.4%	12.0%	34.0%	75.0%	90.0%
2017	Deloitte	3.0%	16.8%	60.0%	–	–
	Average EV market share	2.9%	11.5%	29.5%	69.5%	85%

Source: Own representation based on Bank of America Merrill Lynch (2017), Bank of America Merrill Lynch (2018), Morgan Stanley (2017), UBS (2017), J.P. Morgan (2017), Deloitte (2017a, b)

Table 3.2 Market share EV sales in million units

Release date	Model/Study	2020	2025	2030	2040	2050
September 2017	Morgan Stanley		10.00			108.00
November 2017	UBS		16.20			
November 2017	JP Morgan	1.90	4.00	9.00		
March 2018	Bank of America Merrill Lynch	2.38	13.56	40.45	89.25	107.10
2017	Deloitte	–	–	–	–	–
	Average EV market share	2.14	10.94	24.73	89.25	107.55

Source: Own representation based on Bank of America Merrill Lynch (2017), Bank of America Merrill Lynch (2018), Morgan Stanley (2017), UBS (2017), J.P. Morgan (2017), Deloitte (2017a, b)

Table 3.3 Market share ICE sales

Release date	Model/Study	2020	2025	2030	2040	2050
September 2017	Morgan Stanley	96.0%	91.0%	84.0%	37.0%	20.0%
November 2017	UBS		84.2%			
November 2017	JP Morgan	98.0%	96.2%	92.0%		
March 2018	Bank of America Merrill Lynch	97.6%	88.0%	66.0%	25.0%	10.0%
2017	Deloitte	97.0%	83.2%	40.0%		
	Average EV market share	97.2%	88.5%	70.5%	31%	15%

Source: Own representation based on Bank of America Merrill Lynch (2017), Bank of America Merrill Lynch (2018), Morgan Stanley (2017), UBS (2017), J.P. Morgan (2017), Deloitte (2017a, b)

Table 3.4 Average EV and ICE market share in million units

Release date	2020	2025	2030	2040	2050
Average ICE market share	71.73	84.2	59.1	39.8	18.97
Average EV market share	2.14	10.94	24.73	89.25	107.55

Source: Own representation based on Bank of America Merrill Lynch (2017), Bank of America Merrill Lynch (2018), Morgan Stanley (2017), UBS (2017), J.P. Morgan (2017), Deloitte (2017a, b)

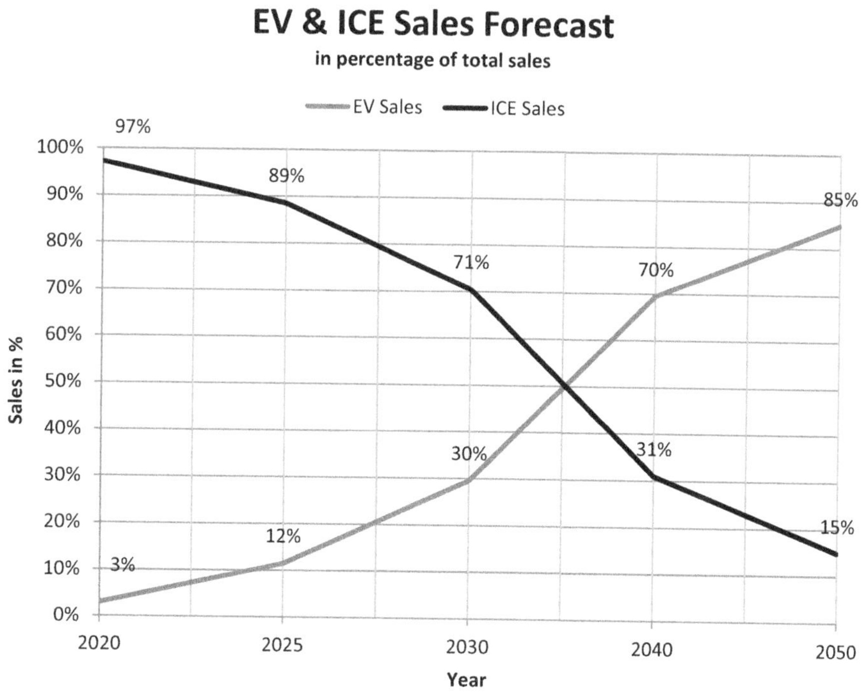

Fig. 3.1 Global EV and ICE Sales Forecast—2050. Source: Own representation

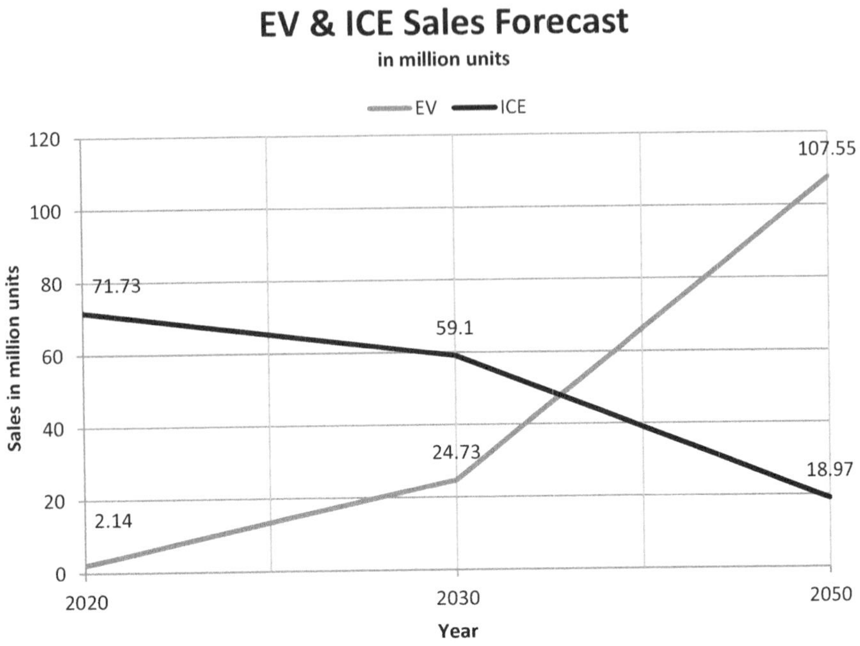

Fig. 3.2 EV and ICE Sales Forecast in million units—2050. Source: Own representation

7 trillion Dollars industry by 2030. Major drivers for this big market potential are e-mobility, shared mobility, and digitalization (McKinsey, 2016). Bringing this result into relation with the developed forecast model, the automobile industry will reach a revenue potential of more than 30 Trillion Dollars by 2050.[1]

This calculation assumes that the potential revenue is correlating with the market share of EVs, as it is one key driver of the revenue of the future automobile industry. However, additional variables have to be analyzed in order to investigate the future potential revenue of the automobile industry. Due to limitations of this research, this investigation will not be longer pursued further within this research, as it requires in-depth information which cannot be accessed by this time. Additionally, the future automobile market is shaped by uncertainties such as the development of customer preferences, which would impair the reliability and recoverability of the analysis.

[1] If 24.73 million EVs in 2030 equal 7 trillion Dollar s revenue, then 107.55 million EVs by 2050 equal 30.44 trillion Dollars revenue.

References

Allgemeiner Deutscher Automobil-Club e.V. (2018, April). *Was kosten die neuen Antriebsformen?* Retrieved September 20, 2021, from ADAC Website: https://www.adac.de/_mmm/pdf/E-AutosVergleich_260562.pdf

Bank of America Merill Lynch. (2017). *Global electric vehicle primer: Fully charged by 2050.* Bank of America Merill Lynch, Automobile Industry.

Bank of America Merrill Lynch. (2018). *Global electric vehicles: A look at what changed in 2017.* Bank of America Merrill Lynch, Automobile Industry.

Bundesministerium für Wirtschaft und Energie. (2014). *Rahmenbedingungen und Anreize für Elektrofahrzeuge und Ladeinfrastruktur.* Retrieved September 20, 2021, from BMWi Website: https://www.bmwi.de/Redaktion/DE/Artikel/Industrie/rahmenbedingungen-und-anreize-fuer-elektrofahrzeuge.html

Center of Automotive Management. (2017). *AutomotiveINNOVATIONS 2017.* Retrieved September 20, 2021, from Autoinstitut Website: http://auto-institut.de/innovations_studien.htm

Deloitte. (2017a). *Mobilität 2.0.* Deloitte, Automotive intellience; Automobile Industry.

Deloitte. (2017b). *The future of the automotive value chain.* Retrieved September 20, 2021, from Deloitte Website: https://www2.deloitte.com/content/dam/Deloitte/de/Documents/consumer-business/Deloitte_Sonderbeilage%20Automobilwoche2017_safe.pdf

J.P. Morgan. (2017). *EVs: Expectations versus reality.* J.P. Morgan.

McKinsey & Company. (2016, February). *Beyond the supercyle.* Retrieved September 20, 2021, from McKinsey & Company Website: https://www.mckinsey.de/files/170216_pm_mgi_beyond_the_supercycle.pdf

McKinsey & Company. (2017). *E-mobility.* Retrieved September 20, 2021, from McKinsey & Company Website: https://www.mckinsey.de/files/170104_pm_e-mobility.pdf

Morgan Stanley. (2017). *Electric vehicles: On the charge.* Morgan Stanley, Automobile Industry.

Schweitzer, H. (2018, March 19). *Diese Elektroautos kommen jetzt.* Retrieved September 20, 2021, from Zeit Online Website: https://www.zeit.de/mobilitaet/2018-03/elektromobilitaet-elektroautos-kaufpraemie-modelle-2018-2019

Tesla. (2014). *Tesla Gigafactory.* Retrieved September 20, 2021, from Tesla Website: https://www.tesla.com/de_DE/gigafactory

The Boston Consulting Group. (2017, November 6). *Elektrofahrzeuge werden ab 2030 voraussichtlich die Hälfte des weltweiten Automobilmarktes ausmachen.* Retrieved September 20, 2021, from The Boston Consulting Group: https://www.bcg.com/de-de/d/press/06Nov2017-PM_The-Electric-Car-Tipping-Point-175834

UBS. (2017). *Global Autos/UBS Evidence Lab.* Automobile Industry.

Viehmann, S. (2018, January 19). *Elektroautos.* Retrieved September 20, 2021, from Focus Online: https://www.focus.de/auto/elektroauto/marktzahlen-2017-elektroautos-deutschland-verdoppelt-marktanteil-aber-stromer-absturz-in-den-niederlanden_id_8329020.html

Chapter 4
The Impact of E-Mobility on the Automobile Industry

With new powertrains, new mobility concepts and adapting charging infrastructure the value chains of traditional automotive manufacturers will change.

What exactly does this change include specifically? What are differences between the production of an EV and which other differences occur between traditional cars and EVs? These questions will be answered in the first section of this chapter.

The next section identifies how e-mobility is transforming the traditional concept of mobility, because e-mobility does not come only with electric engines, it also comes with shared economies and other trends.

The last section will deal with the charging infrastructure and how it will change. It will mainly deal with the question "How will the infrastructure look like for the era of e-mobility?"

The analysis also shows that the following three major changes will be impacting the automobile industry: Transition from ICEs to electric engines, implementation of a new concept of mobility, and charging electric infrastructure.

The traditional value chain of OEMs was based on ICE production. Due to the implementations of EVs within the value chain it has been analyzed how components of EVs differ from traditional ICEs. As a result, EVs cause an elimination of traditional components (e.g., the ICE itself, exhaust system or fuel system, injection system, fuel tank and pump, etc.) and require modified components (e.g., the gear, heat regulation in the interior, bodywork) as well as new components (e.g., electric motor with electronic control and suitable energy storage, new heating system, braking system, etc.).

As new mobility concepts will rise, shared economy is causing a big cultural shift. Car sharing services such as Uber are changing behavior and cultural values in the long term and change the traditional mindset of getting from point A to B.

Furthermore, the necessary infrastructure provides opportunities for companies in different sectors to develop new and innovative business models and to benefit significantly from the introduction of e-mobility.

K. Y. Ciftci et al., *The Potential Impact of E-Mobility on the Automotive Value Chain*, SpringerBriefs in Business, https://doi.org/10.1007/978-3-030-95599-1_4

OEMs can extend their value chain by provisioning charging infrastructure and e-mobility services. This allows OEMs to become a full-service provider for the infrastructure besides their provision of EVs.

The entrance of OEMs into this market carries many risks and is far from their current core business. However, OEMs can satisfy customer preferences and therefore increase demand for EVs.

4.1 The Impact of E-Mobility on the Automobile Industry

There are a lot of changing issues because of e-mobility in the market. Due to limitations, this research will focus on changes which are directly affecting the value chain of large automotive manufacturers.

There are three major changes/shifts which are caused by e-mobility and have a significant impact on the automotive value chain:

1. New powertrains
2. Era of new mobility
3. Charging infrastructure

Additional e-mobility has an impact on the employment of the automobile industry. However, the issue of employment will not be addressed within this chapter, because it is not significantly affecting the automotive value chain.

4.2 Electrification

As described within the previous chapter, EVs will definitely enter the market due to driving forces which has been also analyzed in Chap. 3. Several studies and forecast models of consulting firms and banks have shown the development of EVs and that by 2050, almost every sold car is an electric car (Siegfried, 2021a).

In the next section, it will be identified how e-mobility is transforming the traditional concept of mobility, because e-mobility does not come only with electric engines. It also comes with shared economies and other trends.

All identified changes in this chapter will lay the basis for the following chapter. This chapter deals with the external changes due to EVs and what the circumstances (action) will be like within the world of EVs. The next chapter will deal with the OEMs and how they can react (reaction) to these circumstances in order to maintain their powerful position within the automotive industry.

The conversion to e-mobility in particular is linked to with challenges that cannot be solved with the traditional processes and production lines. At the moment, the main focus of OEMs shall, therefore, be on the integration of EVs into their existing production systems.

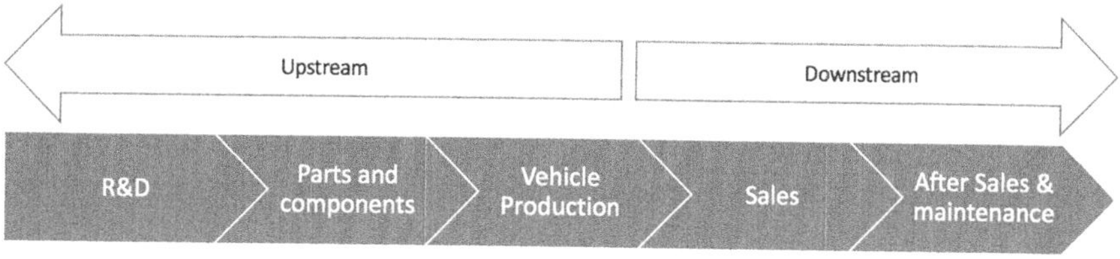

Fig. 4.1 Traditional automotive value chain. Source: Own representation based on Mohseni (2018) and Diehlmann and Häcker (2012, pp. 2–6)

While VW sets up its own production facilities for EVs, almost all other manufacturers are currently converting their large production plants.

The integration of EVs leads to an additional complexity. All the modules and systems that make up a car have to be available just-in-time for all models. EVs do not only need new engines and batteries, but also new heater and air-conditioning systems. Porsche, therefore, relies on variable cycle lengths and a flexible production line. For sensitive activities, the new transport system can stop and then move to the next production stage station more quickly.

The complexity of the supply chain can be seen in the example of heating: EVs require electric heating as the combustion engine is eliminated as a heat source. So far this is only produced in very small numbers. In the future, medium-sized heating suppliers will, therefore, have to manufacture large series, which in addition require a high degree of technological know-how (Kallina & Siegfried, 2021).

In order to understand how the supply chain will change it has to be understood how ICEs are produced and then it has to be identified which car components of an ICE will remain in the EV and which complements will be replaced or even eliminated. It has to be compared how the components of an ICE differ from the components which are needed for an EV.

First of all a rough explanation of the traditional automobile value chain will be given (Fig. 4.1).

The "Research and Development" area represents the first step in the automotive value chain. The focus is on developing new vehicles and researching new technologies for existing vehicles. In contrast to many other areas of the value chain, the responsibility for R&D still lies within the OEMs (Siegfried & Zhang, 2021). The "Parts and Components" department which consists of tier 1–3 suppliers, is also an upstream activity and is directly linked to the R&D department.

The department of parts and components is managed of clearly structured supply chains. Raw materials of tier-3 and tier-2 suppliers are further processed to components. Based on these components, tier-1 suppliers finally produce finished modules or systems which will then be delivered to OEMs. These modules, e.g., can be parts of the drive or chassis electronics or the air-conditioning system. In contrast to the development sector, this part of the value chain is most likely to be outsourced to linked suppliers (Mohseni, 2018).

Differentiated Customer preferences are increasing the individualization of a car. This leads to an extensively wide range of varieties and car models. As a result complexity rises and requires a high degree of flexibility from OEMs. Therefore, OEMs are increasingly outsourcing their supply chain activities. Additionally, the product life cycle of cars decreased significantly in the past. This leads to higher cost pressure which forces OEMs to shift their supply chain activities toward suppliers. Today OEMs have around 30% in-house production depth of their cars (Gropp M., 2016).

The manufacturers themselves retain competencies in the field of bodywork, engines, powertrain, and exterior. Since all of these elements essentially embody the brand of the OEMs and thus represent performance and design characteristics, the OEMs try to retain a large part of this value chain (Mohseni, 2018).

Components that are less noticed by the customer and not represent a differentiation characteristic (such as electrics) are almost completely outsourced. The distribution of the produced automobiles to the customers finally takes place through local distribution centers and car dealers such as own distribution companies, independent dealers, or mixed forms (Özelci, 2018). In addition to traditional sales, other forms of financing such as leasing are offered from auto banks to customers. After all, the after sales area is the last part of the automotive value chain. This includes activities such as the workshop and spare parts business. Despite a relatively low share of sales, this part of the added value represents an important subarea for the automobile manufacturers and contributes massively to the profits of an OEM (Book, et al., 2012).

This value chain represents the traditional value chain of OEMs which have formed their value chain based on ICE production. In order to see what will change, it has to be identified how the parts and components of an ICE differ from an EV.

Three different clusters can be formed for the component analysis. The first cluster contains all components and parts which are not needed for EVs. These components will be eliminated for the production of EVs. The second cluster contains all components and parts which have to be modified for the production of EVs. The third cluster contains all components and parts which have to be added new in order to produce electric cars.

4.2.1 Eliminated Components

Beginning with the first cluster which includes parts that are not needed in EVs and fully connected to the internal combustion engine. This applies to main components such as the ICE itself, but also to components that are indirectly responsible for the movement of the vehicle. Functionally connected components of the ICE systems such as the exhaust system or fuel system, consisting of injection system, fuel tank, and pump, are therefore eliminated. Further drastic changes to the powertrain can be seen by looking at the torque. In contrast to the combustion engine, the maximum torque of the electric motor is already available in the low speed range during start-up, thus enabling the enormous initial acceleration. As a result, components such as a clutch or manual transmission will be eliminated in the future for EVs.

An alternator is also no longer required, since the electrical power supply must be ensured by other sources (Özelci, 2018).

By eliminating these components, the added value share of OEMs continues to decline and reduces mainly to the basic concept and the design of the vehicles. This makes the manufacturers vulnerable to newcomers, since they increasingly function as well-organized system integrators of the individual systems. The elimination of these components is also affecting the core business of OEMs. By replacing the ICE with batteries, core competences of OEMs are endangered. The brand identity of OEMs will be challenged to theses eliminations. Car engines are a strong differentiating characteristic of a car brand, which will be challenged due to the elimination of the engine. And also the after sales department will be affected significantly. The ICE consists of 1400 components while the EV consists of 210. This will reduce maintaining activities for OEMs. All these challenges will be discussed within the next chapter. Nevertheless, this should show how challenging eliminating components for OEMs actually are (Klesse, 2011).

4.2.2 *Modified Components*

The components used in modified form for EVs include among others the gear. The EV is an automatic vehicle, which significantly reduces the number of gears compared to the conventional car. So the "BMW i3" or "VW e-up!" have a one-speed gearbox. This is due to the fact that the electric motor immediately transmits the maximum torque to the wheels, making individual gear stages unnecessary. Even electric cars with high maximum speed such as the "Tesla S" pass on the torque of the electric motor to the rear wheels only by means of a one-speed gearbox and therefore reach top speed of up to 210 km/h. This reduction in gears leads to a significant reduction in complexity of the automobile (Aradex AG, 2018; Kane, 2015).

An additional modified component is the heat regulation in the interior of the cars. While in the ICE, the heat of the engine can be used for internal heating, the EV requires a separate heater which will run by electricity. The same applies on cooling and air-conditioning compressors (Drage et al., 2017, pp. 40–46).

Another modification is the bodywork. The heavy engine block, gearbox, and numerous other components would make electric vehicles actually lighter because these components will be eliminated with the EV. However, this is not the case, because the energy density of the battery cells compared to liquid fuel is much lower.

Battery packs are significantly heavier than the conventional powertrain and cause nearly 20% of the curb weight in an average midsize electric car. In premium EVs, which have significantly higher ranges and maximum speeds, this value is even around 36%. The fact that the battery accounts for a large proportion of the total weight of the vehicle leads to improvements in the bodywork. Instead of heavy aluminum panels, therefore, carbon fibers are used for EVs (Mohseni, 2018).

4.2.3 New Components

The first and most obvious substitution results from the replacement of the ICE. The classic internal combustion engine as well as the associated exhaust system and the fuel tank is removed and replaced by an electric motor with electronic control and suitable energy storage in the form of a traction battery. The further development of the above-mentioned wheel hub motors is strongly emphasized in the course of electrification. In addition to the obvious package advantages, wheel hub motors offer the opportunity to remove the classic powertrain with transmission, differential, and Cardan shaft or drive shaft and, therefore, improve efficiency by eliminating the various translations and the friction losses.

The energy required to provide the drive power is no longer carried in battery electric vehicles in the form of fuel in the tank vessel, but stored exclusively in a battery. Battery systems based on lithium-ion technology are currently suitable for the use as traction batteries in EVs. This is due to the fact that lithium-ion batteries offer very high energy densities. As a result, more energy can be carried in the same weight compared to other battery systems, which has a positive effect on the achievable range. Lithium-ion technology has evolved significantly in recent years, yet its potential for further development is expected.

The electric energy stored in the traction battery cannot be used directly to provide power. The constant DC voltage of the battery must first be converted into the form required for the motor. For DC motors this is done by adjusting the voltage and current to the operating point and the torque required by the driver. For the operation of the frequently used asynchronous or synchronous machine, however, a three-phase alternating is required, which must first be generated by means of a corresponding power electronic. In addition to the control of the motor by the power electronics, a higher-level control unit is used. This processes the data coming from the vehicle sensors and provides functions such as slip control and stability control (Özelci, 2018).

The substitution of the internal combustion engine also has an effect on auxiliary components, which are closely linked to the internal combustion engine. Among other, this affects the thermal management of the vehicle. Its tasks include not only the interior air conditioning but also the cooling of the electric components. The classical tasks are extended in the EV by the temperature management of the battery. When starting the vehicle at low outdoor temperature, a heating of the battery may be required. This is due to the fact that lithium-ion batteries can only deliver very low power at low temperatures. In the past, the loss of heat caused by the internal combustion engine has reliably brought the cooling water to a high temperature. This is no longer the case due to the higher efficiencies of the electrical components.

The cooling water of the engine can now no longer be used as a heat source. Therefore, the heat energy for the air conditioning of the interior must be provided through appropriate systems. This can be done directly with the heating of the indoor air by the so-called PTC heaters or indirectly by the heating of a liquid medium. Basically, the conversion of electrical energy into heat is not difficult. However, the energy required for this has to be taken from the battery and is, therefore, no longer available for the drive. This negatively influences the range of the EV. The air-conditioning compressor required for cooling must be electrically driven directly in the EV (Zhen Zhu, 2014, pp. 14–28).

In the case of the braking system, changes in the electrified powertrain can be expected due to the possibility of recuperation. While in conventional vehicles the conversion of kinetic energy into heat has to be done exclusively by the braking system, in the electrified powertrain the braking power is used as far as possible to drive the generator for the electric motor. The electrical energy generated by the generator recharges the energy storage. Since the required electric maximum braking power for safety reasons usually exceeds the drive power, conventional braking systems must still be used. These support the electric motor when braking, if necessary. The distribution of the braking force is called brake blending. The brake blending requires that the conventional part of the brake system, in contrast to today's usual mechanical control by an electrical control. In any case, there has to be an intelligent link between electric braking and pedal actuation for the mechanical brake. In EVs the event of braking does not come directly from the pedal actuation, but from the electronic system (Bosch, 2018).

Steering is another approach to change as the drivetrain is electrified. The classic hydraulic steering assist is based on a belt-powered pump that provides hydraulic pressure for power steering assistance.

In electromechanical power steering, the steering movement is assisted by an electric motor acting on the steering column or on the rack.

In a conventional steering system the rotation that the driver performs on the steering wheel is mechanically transmitted to the wheels. EV producers avoided the direct connection between the steering wheel and wheels (Bosch, 2018). The steering movement is converted into electronic signals, which are transmitted to a control unit. This in return sends commands to actuators on the front axle, which then steer the wheels. As a result, the intention of the driver is transmitted even faster to the wheels than with a conventional mechanical steering. This system is called steer-by-wire and is already used by EV manufacturers like Nissan and Tesla already (Welt, 2013).

This steering concept is also the basis for autonomous driving cars. Steering assistant systems will be strongly developed in order to use these systems for autonomous driving. The car supplier Bosch merged in 2015 with ZF (Freidrichshausen Lenksysteme GmbH), which only focused on developing intelligent steering systems for EVs. This shows how important steering assistant systems will be for EVs and future autonomous driving. Bosch's named their company "Robert Bosch Automotive Steering" (Bosh, 2018).

Bosch CEO Volkmar Denner believes that electric power steering is the basic technology for the most important trends in the automotive industry: autonomous driving and energy-efficient cars such as EVs (Frankfurter Allgemeine Zeitung, 2014).

4.3 New Concept of Mobility: Shared Economy

Getting from point A to point B is going to be transformed significantly.

Autonomous vehicles, Electrification, and shared ownership of vehicles will become the 3 key triggers for the transformation of mobility.

As in Chap. 3 already discussed people are moving into cities and moving away from rural areas. And this trend is continued to move up. Due to moving populations into cities and private car ownership increase the CO_2 pollution. Populations are wasting very much time for searching parking spaces. Transportation takes around 60% of the overall energy use in cities. In urban cities around 40% of total gasoline use is caused by cars looking for parking space.

Public transportation is not solving these problems due to the first mile last mile problem. The first mile last mile problem addresses the movement of people from a hub to a destination. It deals with the problem "how someone does gets from their final/beginning destination to the final destination." The beginning/final destination may be far away from the public transportation. Therefore, the emergence of vehicle sharing is rising, which is tremendously decreasing the amount of congestion and the amount of cars. Also bicycle sharing is rising. Basically vehicle sharing models enable people to pick up vehicles and drop of cars wherever they want. By allowing people to use vehicles without bringing the vehicle to the same place where they picked up the vehicle, solves the first mile last mile problem. This creates a connection between public transportation and public transportation without having the first mile last mile problem (Donelly, 2015).

People's relationships with cars are chasing. Ownership has changed to renting. Car sharing is the result of the change in perception and behavior of transportation.

There are already well-established models and apps which are shaping the mobility landscape already. Car2go, Moovel, and Mytaxi are on-demand flexible car sharing models of Mercedes-Benz. Uber and Lyft are also well-known models for ride sharing (Daimler, 2018).

Generally, there are four types of car sharing. The largest proportions of all users have station-based and "free floating" car sharing (Dieringer, 2016, pp. 6–18).

Classic Car Sharing (Stationary)

The car is picked up at a fixed station and returned at the end of the rental period at the same or another defined location. This has the advantage that you do not have to search for the car as it can happen during "free floating" (for example, from Car2go, DriveNow). On the other hand, it includes the first mile last mile problem, where people have to manage their way to the vehicle. It may be that the vehicle is not located close to their location. The cars are distributed throughout the business area, can be searched and booked by app or website (Daimler, 2018).

As the name "stationary car-sharing" implies, there are fixed stations. This is the oldest variant of car sharing. Someone books the vehicle in advance on the Internet or via smartphone app. The user has to select the desired station and one of the local vehicles. The duration of the booking is generally determined beforehand.

The advantage of this variant is the low cost per hour. Most providers in Germany have been relying on this model for many years. There are many vehicle classes available for rent (Deloitte, 2016a, b).

Flexible Car Sharing

Flexible car sharing is also known as "free-floating" car sharing or "station-independent car sharing." This type of car sharing is mainly available in big cities. There is a solid business district in the city. Within this area people can pick up and drop off the vehicles anywhere without being tied to stations. The cars distribute themselves through the rides of the customers themselves. This concept eliminates the problem of last mile, because the driver can drop the car off at his final desired destination.

If someone happens to see a vehicle, he can open the vehicle by smart card or smartphone and drive. A reservation is not necessary. Alternatively, people can look on the Internet or by app, which vehicles are available in the local area. These can be reserved for a short period before the rental (Deloitte, 2016a, b).

The advantage of this variant is the very high flexibility. The disadvantage is the high costs for longer trips. It is billed per minute or kilometer. Above all, the automakers are using their car sharing offerings on this relatively new model. More and more providers are also offering a combination of stationary and flexible car sharing.

Private Car Sharing

In addition to the professional car sharing of large providers with their own fleets, there are also more and more private individuals who provide their car temporarily to other private persons. For this purpose, there are websites that specialize in the mediation of tenants and car owners. This is an interesting alternative, especially for rural areas.

Advantage of this variant is the low price and the large vehicle diversity. A disadvantage is that the vehicles are partly older and are not checked by the agents. The insurance usually runs through the intermediary (Mattke, 2018).

New Shared Taxis

Unlike conventional taxis Uber cars are not recognized as "taxis." The cars have no fixed stands and no Uber sign on the roof. The communication between customers and drivers comes about exclusively via the smartphone app. The app is extremely easy to use. The registered user gets the information where the Uber car is currently located, and also will be informed how long the journey is likely to take and what it will cost. The user will be billed cashless on the previously registered credit card. 20 percent of the fare is paid to Uber, the rest is received by the driver. This is how Uber and other car sharing providers like Lyft work (Deloitte, 2016a, b).

Car sharing services are changing behavior and cultural values.

Shared economy is causing a big cultural shift. Car sharing models like Uber are a part of the new concept of mobility. The idea of entering a strangers car to get to a destination is a totally new perception which did not exist before. In the past it has been a cultural principal not to enter strangers cars because it is too dangerous not knowing the driver. Now people are paying a premium to get inside a strangers' car.

Parents are trusting these models and letting their children go into a car every day without knowing who the driver is. This shows how car sharing is actually changing cultural values (Vaynerchuck, 2017).

But people trust the power of sharing because Uber, Lyft and co. are not selling the cultural shift but the idea of new mobility. They are convincing people that the new mobility concept is the future (Siegfried, 2021b). Basically people do not like change. However, the benefits and the philosophy of car sharing models are making people ready for change. And, therefore, Uber, etc., is selling a philosophy. The philosophy presents a sustainable and responsible world where transport is accessible for everyone, everywhere, and every time.

4.4 Charging Infrastructure

As already described, infrastructure is a crucial factor for EV demand. Customers think it is important to have a broad EV charging infrastructure, to recharge their EVs fast and immediately. This has been discussed in the section of "customer preferences" in Chap. 3. This chapter will explain the different charging types and how they differ from each other considering costs and charging time.

For recharging EVs, there are currently three types of charging stations available in 2018:

1. *Smart charging station*
   Private wall box. These are often used in the private sector. However, charging less than 3.7 kW takes time.
2. *Standard charging station*
   Normal charging points with a maximum charging power of 22 kW, which have often been used in public or city areas.
3. *Fast charging station*
   Fast charging points with a charging capacity of more than 22 kW. These include the so-called superchargers, so high-performance DC charging stations with very short charging times and a charging power of 120 kW or more. So far the common problem is that the vehicle and its charging technology must be designed for the charging power, which is largely not the case with the currently offered electric vehicles.

These three different types also have different costs of usage, costs of installation, and charging durations.

Smart charging stations' installation costs amount around 2700 euros and running costs of around 500 euros each year. In order to charge an EV, taking a Renault Zoe as reference model, the charging duration would take 11 h.

Standard charging stations with 22 kW power have installation costs of around 7500 euros and running costs of around 750 euros per year. A full charge of a Renault Zoe would take 1.9 h with a 22 kW station.

A fast charging station includes initial installation costs of 25,000 euros and 1500 euros running costs per year. To charge a Renault Zoe with a 50 kW station, would take only 0.8 h. A supercharger with 120 kW power is not able to charge a Renault Zoe. However, it is able to use supercharge stations with 120 kW with the Tesla model S, which would take 0.8 h.

In order to charge a Tesla model S with a smart charging station would take 27 h, with a standard charging station, 4.6 h, and with a fast charging station (50 kW) 2 h (Deloitte, 2018).

The additional infrastructure provides opportunities for companies in different sectors to develop new and innovative business models and to benefit significantly from the introduction of e-mobility. For sustainable and successful investments in the new technology and the construction of the charging infrastructure, companies have to take into account a variety of economic, technical, and sociological factors.

OEMs with corresponding core competencies and access to the public can expand their existing business model with the new developing infrastructure. They can extend their value chain by provisioning charging infrastructure and e-mobility services. This allows OEMs to become a full-service provider for the infrastructure besides their provision of EVs.

The infrastructure industry is in a very competitive situation. Infrastructure development can be targeted by former energy suppliers and also by previous infrastructure operators (municipal utilities, grid operators, etc.). The entrance of OEMs into this market is very risky, and far from their current core business, however OEMs can satisfy customer preferences and therefore increase demand for EVs. Additionally, OEMs would be independent from the infrastructure industry and could create their own infrastructure framework (Bozem & Rennhak, 2013, pp. 210–221).

The question, how OEMs can use the developing charging infrastructure to add value to their current business model, will be discussed in the following chapter.

References

Aradex AG. (2018). *Gearbox vs. direct drive*. Retrieved September 20, 2021, from Aradex website: https://www.aradex.de/en/system-solutions/whitepaper-power-electronics/gearboxes-vs-direct-drive/

Book, M., Ellul, E., Ernst, C., Frowein, B., Kreid, E., Rilo, R., et al. (2012). *The European automotive aftermarket landscape*. The Boston Consulting Group, Automobile Industry, Market Dynamics.

Bosch. (2018). *Hydraulic and electrohydraulic steering systems*. Retrieved September 20, 2021, from Bosh Website: https://www.bosch-mobility-solutions.com/en/products-and-services/commercial-vehicles/steering-systems/hydraulic-and-electrohydraulic-steering-systems/eshydraulicpressureforpowersteeringassistance

Bosh. (2018). *Berlin-Robert Bosch Automotive Steering GmbH*. Retrieved September 20, 2021, from Bosh Website: https://www.bosch.de/unser-unternehmen/bosch-in-deutschland/berlin-robert-bosch-automotive-steering-gmbh/

Bozem, K. N., & Rennhak, C. (2013). *Energie für nachhaltige Mobilität*. Springer Verlag.

Daimler. (2018). *BMW Group and Daimler AG combine mobility services*. Retrieved September 20, 2021, from Daimler Website: https://www.daimler.com/case/shared-and-services/en/

Deloitte. (2016a, July). *Car sharing in Europe*. Retrieved September 20, 2021, from Deloitte Website: https://www2.deloitte.com/content/dam/Deloitte/de/Documents/consumer-industrial-products/CIP-Automotive-Car-Sharing-in-Europe.pdf

Deloitte. (2016b, June). *Car sharing in Europe*. Retrieved September 20, 2021, from Deloitte Website: https://www2.deloitte.com/content/dam/Deloitte/de/Documents/consumer-industrial-products/CIP-Automotive-Car-Sharing-in-Europe.pdf

Deloitte. (2018, March). *E-mobility: Ladeinfrastruktur als Geschäftsfeld*. Retrieved September 20, 2021, from Deloitte Website: https://www2.deloitte.com/content/dam/Deloitte/de/Documents/risk/Risk-Deloitte-Ladeinfrastruktur.pdf

Diehlmann, J., & Häcker, J. (2012). *Automobilmanagement: Die Automobilhersteller im Jahre 2020* (2nd ed.). Walter de Gruyter.

Dieringer, T. (2016, September 22). *Car sharing Flexibilization: Station-based and Book*. (U. Hannover, Ed.). Hannover. Retrieved September 20, 2021, from https://www.iwi.uni-hannover.de/fileadmin/wirtschaftsinformatik/Abschlussarbeiten/MA_Dieringer_T._K..pdf

Donelly, B. (2015, November 16). *How ridesharing could help solve the last-mile problem*. Retrieved September 20, 2021, from mobility lab website: https://mobilitylab.org/2015/11/16/ridesharing-the-last-mile/

Drage, P., Seebald, F., Paul, C., & Hinteregger, M. (2017, September). Innovative HVAC concepts for future vehicles. *ATZ Worldwide* (09/2017) (pp. 40–48).

Frankfurter Allgemeine Zeitung. (2014, September 15). *Bosch übernimmt ZF Lenksysteme komplett*. Retrieved September 20, 2021, from Frankfurter Allgemeine Website: http://www.faz.net/aktuell/wirtschaft/autozulieferer-bosch-uebernimmt-zf-lenksysteme-komplett-13154606.html

Gropp, M. (2016, August 24). *Verbunden auf Gedeih und Verderb*. Retrieved September 20, 2021, from Frankfurter Allgemeine: http://www.faz.net/aktuell/wirtschaft/unternehmen/autohersteller-und-zulieferer-verbunden-auf-gedeih-und-verderb-14402472.html

Kallina, D., & Siegfried, P. (2021). Optimization of supply chain network using genetic algorithms based on bill of materials. *The International Journal of Engineering & Science*. ISSN: 2319-1813. https://doi.org/10.9790/1813-1007013747

Kane, M. (2015, September 22). *Test results show 18% reduction in energy consumption for 3 speed versus 1 speed EV transmission*. Retrieved September 20, 2021, from inside EVs website: https://insideevs.com/test-results-show-18-reduction-energy-consumption-3-speed-versus-1-speed-ev-transmission/

Klesse, H. (2011, September 14). *Autohersteller büßen Macht ein*. Retrieved September 20, 2021, from Wirtschafts Woche: https://www.wiwo.de/technologie/mobilitaet/elektroautos-autohersteller-buessen-macht-ein-seite-2/5212332-2.html

Mattke, S. (2018, January). *Sharing Economy: Plattformen für privates Carsharing werden zu gewerblichen Marktplätzen*. Retrieved September 20, 2021, from Heise Website: https://www.heise.de/newsticker/meldung/Sharing-Economy-Plattformen-fuer-privates-Carsharing-werden-zu-gewerblichen-Marktplaetzen-3919715.html

Mohseni, A. (2018, June 4). *Head of business development, continental AG* (K. Ciftci, Interviewer, & C. Kaan, Translator) Escborn, Hessen.

Özelci, M. (2018, May 3). *CEO and owner - Audi car dealer in Frankfurt* (K. Ciftci, Interviewer).

Siegfried, P. (2021a). *Business management case studies*. ISBN: 978-3-75431-691-7. BoD Book on Demand Publisher.

Siegfried, P. (2021b). *Enterprise management automobile industry business cases*. ISBN: 9783753444871, BoD Book on Demand Publisher.

Siegfried, P., & Zhang, J. (2021). Developing a sustainable concept for the urban lastmile delivery. *Open Journal of Business and Management (OJBM)*. https://doi.org/10.4236/ojbm.2021.91015

Vaynerchuck, G. (2017, December 12). *We live in the biggest cultural shift in human history.*

Welt. (2013, December 17). *Proleme mit der Kabellenkung*. Retrieved September 20, 2021, from Welt Website: https://www.welt.de/motor/news/article123028421/Drive-by-wire-im-Infiniti-Q50.html

Zhen Zhu, J. (2014). *Battery thermal management systems of.* Retrieved September 20, 2021, from Chalmers Publications Website: http://publications.lib.chalmers.se/records/fulltext/200046/200046.pdf

Chapter 5
New Entrants

Additionally, three major trends (1. Electric engines, 2. Shared economy, and 3. Autonomous driving and Artificial Intelligence) are opening up markets for new entrants and are consequently shifting OEMs market power which provides new opportunities for companies to enter the market (Michel & Siegfried, 2021).

Following four new entrants have been identified, which are causing a big transition in the automotive value chain: Tech companies, mobility providers, suppliers, and energy providers.

5.1 New Entrants

In the previous chapters, it has been analyzed which influencing factors are driving e-mobility, how e-mobility will develop in the future, and what changes within the automobile industry occur due to e-mobility.

Summarized the following three major changes have been identified for the automobile industry:

1. Transition from ICEs to electric engines
2. Implementation of a new concept of mobility
3. Charging electric infrastructure

The findings from the previous chapters have shown that the automobile industry is underlying a tremendous change.

This chapter will outline who the new entrants of the automobile industry are and which new concept and services they are providing to the industry.

Based on this, the next chapter will then investigate the rising question "How OEMs can adapt their value chain and business concept in order to be resistant against new entrants who are shifting the traditional automobile value chain?"

There are four new entrants which are causing a big transition in the automotive value chain.

K. Y. Ciftci et al., *The Potential Impact of E-Mobility on the Automotive Value Chain*, SpringerBriefs in Business, https://doi.org/10.1007/978-3-030-95599-1_5

1. OEM (traditional value chain)
2. Tech industry (focusing on connectivity, autonomous driving, and digital networking of the future car)
3. Mobility providers (focusing on the provision of cars and contributing to the concept of shared economy and other car sharing concepts)
4. Suppliers (Battery producers who are focusing on the battery production and new engine production)
5. Energy providers (who are focusing on the infrastructure and energy supply of the future cars)

All these new entrants are stretching and shifting the traditional value chain of the automobile manufacturers, by crossing their usual industry borders. Every new entrant is focusing on a specific part of the automotive value chain or adding new parts in it. As a result a new reformed value chain arises.

In the following section, the new entrants will be explained very briefly and show how they are adding services and values to the traditional automobile. This section should only give a quick overview about new entrants. Additional information about these entrants will be outlined in Chap. 7.

Basically there are three major trends which are shifting the industry power of OEMs:

1. Electric engines which are directly affecting OEMs, battery suppliers and suppliers and energy industry.
2. Shared economy which is triggered by mobility providers (e.g., Uber, Lyft, Car2go, DriveNow).
3. Autonomous driving and Artificial Intelligence which are affecting tech companies (e.g., Google).

The three major trends are opening up markets for new entrants and are, consequently, shifting OEMs market power. New entrants such as Google, Apple, or also Tesla are crossing their traditional industry boundaries. As a result traditional OEMs like Daimler, BMW, and VW are losing market power, because new entrants are focusing on special parts of the value chain and increase competitive pressure (Bormann, et al., 2018).

As already mentioned there are several new potential market participants, which are threatening OEMs with increasingly competitive pressure.

5.2 Technology Companies

The ICE has been the primary differentiator between OEMs, which is now replaced by electric engines. Thus, OEMs have to find ways for creating new differentiating factors. Due to new entrants who are going beyond their traditional industry borders, other new differentiation features are coming up. According to a study of the Friedrich-Ebert-Stiftung it will be the digital networking and mobility services which distinguish the automakers of the future. The digital connectivity provides the basis for new services and operating concepts such as autonomous driving. This will cause the extension of the automotive and automotive-related value chain. The traditional production of vehicles will earn less money in the future. Simultaneously, the IT industry, especially companies from Silicon Valley and China enhance the automobile industry with new concepts of driving. New tech companies are aggressively attacking OEMs with their digital competence. They are heavily investing and researching in artificial intelligence which is the basis for autonomous driving and voice controlled interactions between driver and car. This shows how important "IT" will be in the future operations of OEMs. It is a new differentiator for car makers in the future. If OEMs will not change their traditional business concept, they are likely to become "hardware providers" for tech companies, which are adding their digital services to the car. In this case OEMs become suppliers for tech companies.

AI becomes a competitive factor that allows OEMs to differentiate from other OEMs. This applies to the use of AI for autonomous driving as well as for the development of new mobility concepts based on AI. According to a study of McKinsey "Artificial intelligence—automotive's new value chain creasing engine," nearly 70% of customers are ready to change brands for better assisted and autonomous driving features. This is why IT and AI are crucial for OEMs to consider into their future value chain (McKinsey & Company, 2018).

5.3 Mobility Providers

In addition, there are other quite aggressive entrants. Companies such as Lyft or Uber do not rely on the development of vehicles, but want to establish a new mobility concept which lays the basis for new digital networking and mobility services in the automobile industry.

This new entrants have already been discussed within the previous chapter. Additionally, it can be said that increasing urbanization and changes in the mobility preferences of customers are causing new mobility concepts to rise. Megatrends such as "using instead of owning" increased environmental awareness and the flexibility of mobility. This is forcing OEMs to invest and integrate this concept into their future value chain. Furthermore, besides mobility providers such as car sharing companies, mobility services are an emerging market.

The management consulting firm Horváth and Partners and Fraunhofer IAO analyzed how much car drivers are willing to pay for the value-added services for cars. They identified six defined mobility service categories which are communication, productivity, basic needs, comfort, information, and entertainment and found out that on average the willingness to pay for mobility services lays between 20 and 40 euros per month.

This could lead into a rising market for OEMs in the future. Within the German market an annual turnover of several billion euros for mobility services during the next 10 years is predicted. In Japan and the USA, the willingness to pay is even higher (Goffart & Strassberger, 2015, pp. 21–29).

5.4 Suppliers

The importance of suppliers, especially of battery cell producing suppliers is heavily rising. This can be observed from the case between Grohmann, Tesla, Daimler, and BMW (Edward, 2017).

Grohmann builds automated systems and machines for car manufacture. They also focus on the production of battery cell machines and thus suit highly to the philosophy of Tesla. The US automaker Tesla, which wants to reach an output of 500,000 vehicles this year, took over Grohmann Engineering GmbH about 150 million Dollars in an M&A transaction at the end of 2016 (Simmons & Simmons, 2016).

The acquisition of Grohmann was a strategic move of Tesla in order to acquire the competence of battery cell production. Tesla's strategic acquisition caused the conscious expiration of contracts with long-standing customers of Grohmann with Daimler and BMW. This triggered a conflict between BMW and Daimler, which are direct competitors of Tesla. This shows how important the integration of battery sell suppliers will be in the future (Freytag & Peitsmeier, 2017).

5.5 Energy Providers

Another potential new entrant is the energy industry. As mentioned earlier in this book, the industry trends which are driven by renewable energy, energy efficiency, and the energy transition are forcing companies to adapt their business models as energy provider as well. Energy companies are already preparing their businesses for the future.

For energy companies, e-mobility opens up a new market. The e-mobility services of energy companies could be further developed by the increasing decentralization of the energy market. Energy companies will increasingly take on the role of the service provider who will take over the development, operation, and network integration of their private generation facilities for their customers.

Additional information of the new entrants will be found within the following chapter.

References

Bormann, R., Fink, P., Holzapfel, H., Rammler, R., Sauter-Servaes, T., Tiemann, H., et al. (2018, March). *Die Zukunft der deutschen Automobilindustrie*. Retrieved September 20, 2021, from WISO Website: http://library.fes.de/pdf-files/wiso/14086-20180205.pdf

Edward, T. (2017, April 27). *Exclusive: Tesla's Klaus Grohmann ousted after clash with CEO Musk*. Retrieved September 20, 2021, from Reuters website: https://www.reuters.com/article/us-tesla-germany-exclusive-idUSKBN17T2IY

Freytag, B., & Peitsmeier, H. (2017, May 5). *Warum Grohmann für die Autohersteller so wichtig ist*. Retrieved September 20, 2021, from Frankfurter Allgemeine Website: http://www.faz.net/aktuell/wirtschaft/unternehmen/warum-zulieferer-grohmann-fuer-die-autohersteller-so-wichtig-ist-15001500.html

Goffart, K., & Strassberger, M. (2015). Automobile Mehrwertdienste durch Virtuelle Marktplätze. In C. Linnhoff-Popien, M. Zaddach, & A. Grahl (Eds.), *Marktplätze im Umbruch* (pp. 21–33). Springer Verlag.

McKinsey & Company. (2018, January). *Artificial IntelligenceE – Automotives new value creating engine*. Retrieved September 20, 2021, from McKinsey & Company: https://www.mckinsey.de/~/media/McKinsey/Locations/Europe%20and%20Middle%20East/Deutschland/News/Presse/2018/2018-01-04/mckinsey_ai_report_final_january_2018.ashx

Michel, V., & Siegfried, P. (2021). Digitale Speditionen in der Lebensmittellogistik - Digital freight forwarders in food logistics. *Logistics Journal*. https://doi.org/10.2195/lj_NotRev_michel_de_202102_01. ISSN 1860-5923.

Simmons & Simmons. (2016, November 11). *Tesla Acquires Leading German Automation Specialist Grohmann Engineering*. Retrieved September 20, 2021, from Simmons & Simmons Website: http://www.simmons-simmons.com/en/news/2016/november/tesla-acquires-leading-german-automation-specialist-grohmann-engineering

Chapter 6
Value Chain Transition and Potential Strategies

6.1 Value Chain Transition and Potential Strategies

First of all a brief illustration of the hypothesis will be given, in order to get a better understanding of the central question of this chapter. In the following, an Assumption will be formed and further developed into a resulting book and hypothesis, finally the central question will be formed (Brühl, 2011):

Assumption: e-mobility and new mobility concepts will dominate the automobile industry and open up markets for new entrants.

Book: Electric engines and new entrants are offering new concepts and services to the automobile industry and are, therefore, shifting the traditional automobile value chain. (e-mobility impacts the traditional automobile value chain and affects OEMs market power.)

Hypothesis: If OEMs will not react to the value chain shift which is caused by e-mobility and new entrants, OEMs will lose their market power.

Central question:

- How can OEMs change their value chain in order not to lose their market power due to the era of e-mobility?

Based on this central question this chapter is divided into two sections.

Within the first section, the traditional value chain of the traditional OEMs will be presented in order to get an understanding from which basis the new value chain will be redeveloped.

The second section will focus on the new value chain creation.

The future automobile landscape which consists of the changes (Chap. 5) and the new entrants (Chap. 6) will be taken into consideration in order to identify potential markets and strategies for OEMs in the future. Simultaneously, the impact of new entrants and business concepts on the automobile value chain will be outlined.

K. Y. Ciftci et al., *The Potential Impact of E-Mobility on the Automotive Value Chain*, SpringerBriefs in Business, https://doi.org/10.1007/978-3-030-95599-1_6

Additionally, it will be investigated how OEMs can realize opportunities and develop new potential strategies.

6.2 The Traditional Automotive Value Chain

The automobile value chain starts with "Research and Development." The focus is on developing new vehicles and researching new technologies for existing vehicles. In contrast to many other areas of the value added chain, this part of the value chain is settled in the OEM and not outsourced to suppliers. "Parts and Components" is directly linked to the R&D department.

This department acts like an intermediary station between the automobile manufacturers and the suppliers. For the labeling of supplier relationships, the terminology of the automotive industry uses the terms "Tier 1" to "Tier 3". Tier 1 suppliers directly supply to OEMs, whereas Tier 3 supply Tier 2 and Tier 2 provide parts to Tier 1 suppliers (Koch, 2006, pp. 71–84).

Clearly structured supply chain networks arrange the supply of raw materials by Tier 3- and Tier 2-suppliers which are further processed into components. Based on these components, Tier-1 suppliers finally produce finished modules and deliver them to OEMs "Parts and Components" department. These modules can, e.g., be parts for the chassis electronics or the air-conditioning system. In contrast to the R&D department, this value chain station is increasingly dominated by the linked suppliers (Roland Berger, 2017). The reasons for this are increasing customer preferences which force OEMs to offer cars with much higher individualization possibilities (Siegfried, 2014). This pushes the varieties in which cars can be manufactured. This requires more complex production roads. Therefore, the term "mass customization" is used, which means the adaption of car manufacturing in order to stimulate changing customer needs. Simultaneously, this requires high investment engagement and increases variable costs of OEMs. This is why OEMs sourced parts for customer individualization out to suppliers (Talgeri, 2014, pp. 60–68).

The production of parts consists of the assembly and combination of a component with other components. On the next higher level there is the system which is an aggregate of combined components (e.g., the brake system). Modules such as the engine and cockpit, finally, are made by the combination of several systems (Mohseni, 2018).

The distribution of the cars to the customers finally takes place through national distribution subsidiaries and car dealers (Özelci, 2018). In addition to traditional sales, other forms of financing such as commercial banks are also offering financing services such as leasing which are supporting sales volume. Finally, the automotive value chain ends with the department of after sales. This includes activities such as the workshop and spare parts. Despite a relatively low share of sales, this part of the added value represents an important subarea for the automobile manufacturers and contributes massively to the profits (Stricker et al., 2011).

6.3 Transition of the Value Chain and Potential Strategies

Electric engines and its new design cause a profound change of the automotive value chain. The transition of the central core feature, the engine, caused the change of materials, production processes, and quality requirements which apply to every part of the value chain. Additionally, it also shifts the work stake between OEMs and suppliers. Up until now, the core competencies of OEMs and suppliers heavily lied on the ICE. Therefore, e-mobility forces them to adapt their business concepts.

Not only upstream changes such as the production of electric cars, but also only downstream adaption will occupy the car manufacturers.

The energy supply of the EVs and the charging infrastructure will be crucial for OEMs in order to contribute to the initial acceptance of EV in the customer market. Energy suppliers are increasingly turning into charging infrastructure providers and operator. As explained in the previous chapter, industry borders are fading and slowly begin to coalesce.

Another important challenge is the need for innovative distribution concepts and marketing strategies for EVs. The OEM's traditional product-oriented sales strategies such as the traditional distribution by car dealers can no longer be applied to EVs anymore. The reason for that lies in the different mobility concept in which EVs are embedded. New mobility services such as ride-sharing are invented in order to change the traditional inefficient car use. E-mobility forces car companies to act from a customer perspective and change the traditional sales and after-sales strategies into EV applicable strategies (Deutsche Leasing AG, 2018).

These adaptions are linked to challenges for the automobile industry, but also to great opportunities.

OEMs and suppliers, who have their core business in ICEs, lose their differentiation attribute toward their competitors. Traditional core competencies are losing relevance in the industry. OEMs and suppliers have to expand their product portfolios with alternative mobility concepts, and innovative services. According to Madan, partner consultant at KPMG (2018), they need to build up new skills in these sections. Therefore, the allocation of R&D resources in the fields of e-mobility will be crucial.

A new market for components of electric drives such as lithium-ion batteries, electric motors and generators, power electronics and electric systems is rising which concerns the upstream value chain. The part of chemicals and electronics in the automotive value chain is about 30% for an ICE. For EVs the share of chemicals and electronics is 80%. This creates value-added potential for OEMs and automobile companies (Madan, 2018). Downstream opportunities can be found especially in after sales and new mobility concepts. The development and production of EVs requires completely new competencies and thus allows new entrants to operate in the industry. The OEMs core competences are endangered to be replaced by the skills of new entrants.

New players, who were previously not participating in the automotive value chain, have gained significant share of the value chain.

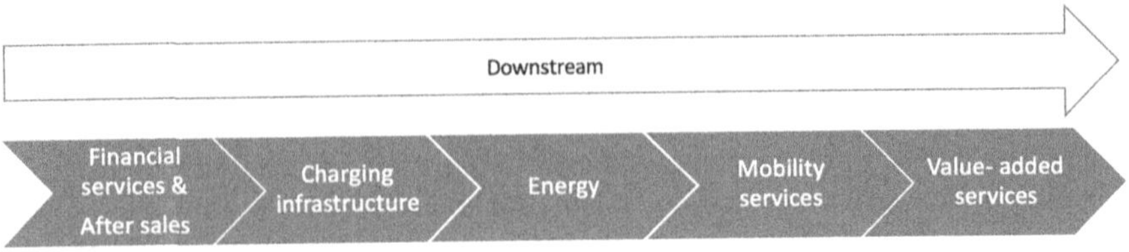

Fig. 6.1 New automotive value chain. Source: Own representation based on Mohseni (2018) and Madan (2018)

The most significant focus lies on downstream activities. Downstream activities have gained the most attention from OEMs, because these activities drive profits significantly and have the highest potential for growth in the perspective of traditional OEMs (Stricker et al., 2011).

Based on the findings and interviews a new electric value chain is presented below (Fig. 6.1):

The new value chain consist of battery manufacturers, electric engine manufacturers, performance electronic companies, other suppliers, manufacturers production, car dealers, financing solutions, charging infrastructure and energy suppliers, and finally mobility and value-added services.

Battery manufacturers, electronics companies, and specialized firms in the upstream sector are getting pressure from new entrants, but also the downstream side will compete with specialized EV dealers, new customer groups such as B2B (e.g., fleet operators), specialized workshops and service providers, charging infrastructure and energy providers. OEM's core business is endangered by new electric vehicle manufacturers like Tesla or BYD (Kamp, 2010).

This is challenging established OEMs in their innovation development and makes them dependent on external competencies.

The development and integration of new components, concepts, and new and external suppliers into the supply chain network will represent a central task for OEMs. Energy supply is likely to emerge as one of the biggest challenge for OEMs to integrate. Because charging infrastructure competence has to be rebuild from scratch due to no previous experience in this industry. However, as it can be observed in the next chapter, infrastructure is not a target section of OEMs.

Batteries account for a huge part of the product differentiation. The battery is determining the performance, weight, and costs of ownership of the EV. Fast charging batteries are thus gaining relevance in the future due to customer preferences which include short charging durations. Consequently, battery technology, therefore, plays a central role for OEMs in the future. The value chain for high-performance batteries is very extensive and includes among other the production of raw materials and materials (metal oxides, metal foils) and the production of anodes, cathodes, electrolytes, and the cell production, the packaging of cells into battery blocks, and integration into the vehicle.

The battery integration consists of the battery packs which have to be linked with the thermal management into the car. The integration of battery packs is largely executed by the OEMs itself. (Lebedeva et al., 2016, pp. 5–11) However, the most value-adding process is the cellular battery production due to its high complexity. It requires very specific know-how and experience to produce batteries cells for EVs (Lebedeva et al., 2016, pp. 26–31).

Market leaders in battery production especially come from Asia.

In 2016, Panasonic represented the largest supplier of batteries for electric cars and held a market share of 40%. BYD, a Chinese former supplier which have developed to an EV manufacturer follows with 20%. LG is held 12% in 2016. Toyota and Panasonic have been producing batteries in a joint venture since 1996 (Wimmelbücker, 2017). Panasonic also works for Tesla. Both have set up the "Gigafactory" in Nevada for the construction of batteries for EVs from Tesla (Hajek et al., 2017).

Summarized it can be said that the need for adapting the automotive value chain for OEMs lies heavily in increasing industry pressure due to new entrants which develop new concepts and technologies. It also has been pointed out that upstream as well as downstream changes have to be undertaken in order to be competitive in the future automobile industry.

In the following, the down and upstream adaptions will be outlined in detail, which OEMs have to take in order to position them in the e-mobility revolution.

6.3.1 Partnerships and Vertical Integration

One strategy for OEMs is to outsource the production of electric engine modules completely or partly to suppliers or other OEMs, who can also provide battery systems. Continental supplies lithium-ion batteries for Daimler and BMW (Rother et al., 2010). Additionally, Continental plans to use battery cells from a joint venture formed this year, with CITIC. (Continental, 2018). Daimler also uses lithium-ion batteries from Tesla and the electric motor from the British Joint Venture Zytec, which has been overtaken by Continental in 2014 (Günnel, 2014). As the head of business development of Continental, Mohseni proved in the expert interview of this book, M&A activities will be a focus of suppliers in order to compete with new entrants (Mohseni, 2018). This shows primary how different the well-structured supply chain network of OEMs and battery supply can be managed. Secondly, it shows the depth of the value creation for EVs.

In order to integrate the new battery supply process into OEMs value chain, they have to decide wisely which suppliers they want to integrate into the value creation process. Additionally, they have to consider the degree between cooperation and protection of know-how. However, by this strategy OEMs are outsourcing components of their value chain and therefore, reduce their value-added activity (Accenture, 2014, pp. 4–7).

Partnerships between Tesla and traditional OEMs highlight the importance of collaborations where OEMs benefit from electric OEMs know-how structure and capital investments. This shows, how new electric OEMs act also as a supplier for traditional OEMs. In cooperation with purely electric car manufacturers, traditional OEMs can gain competencies and product knowledge in the field of e-mobility without developing them independently and within the existing company structures. By this strategy traditional OEMs are avoiding the downside of key investments and risk obligations (Daimler AG, 2014).

Currently the most prevalent partnerships, which can be seen in the area of e-mobility, are collaborations within the battery industry. There have been different forms and structures of partnerships which have established within the automotive industry.

The formation of vertical cooperation can be observed between companies that focus on the same part of the value chain and are competing in the same market.

Successful strategic partnerships follow a clear pattern. The cooperation partner has similar goals and visions of the product which should be created. Such collaborations arise between OEMs and also between suppliers. Partnerships between OEM competitors are resulting from increasing industry pressure and the need to innovate. Additionally, the increasing complexity of innovation is forcing companies to seek for a partner. The collaboration with other competitors aims to increase market power. In addition, it enables companies to develop complex technologic innovations together. Collaborations between competitors are often difficult, as they require cooperation despite competition, but can be used in the area of R&D (Liu & Brody, 2014).

Another form of cooperation is a joint venture. These build up new structures by financial participation of two companies which enable OEMs to be more independent from new start-up and tech companies (Huke & Siegfried, 2021).

A joint venture, on the other hand, refers to the establishment of a joint, legally independent company within the framework of cooperating companies. The participating companies bring different resources and they are involved approximately equally. Joint ventures are also referred to as institutionalized forms of strategic alliances, even though the partners do not necessarily have to operate in the same value chain activities (Boston Consulting Group, 2018).

There are numerous examples for the mentioned collaborations.

After VW and Daimler started to enter the Chinese market for electric cars, in 2010, Daimler entered a joint venture with the Chinese automobile company BYD. The new JV was named Shenzhen BYD Daimler New Technology. BYD was founded in 1995 in Shenzhen and produced only batteries for the consumer sector in the beginning (BYD, 2018). In 2003, BYD started to produce batteries for EVs, which were initially supplied to Chinese electric car manufacturers. The joint venture Shenzhen BYD Daimler New Technology, in which both OEMs hold a 50% share, develops and produces electric cars (Fuchslocher, 2018). Daimler also uses the cooperation with BYD for the access of batteries for their own electric cars. Daimler highly benefits from the experience of BYD with the development and production of electric cars (Krust, 2012).

In 2017, VW entered a JV with the Chinese car manufacturer Anhui Jianghuai Automobile.

In the development and construction of electric cars both companies want to work together in China. For this purpose, a joint venture has been planned. Both groups want to cooperate in research as well as in distribution of the cars and want to offer new mobility services. Anhui J. A. and Volkswagen together want to invest 700 million euros in the production of low-priced EVs (Reuters, 2017).

Jianghuai is one of the largest manufacturers of EVs in China and VW already have joint ventures with the Chinese carmakers First Automotive Works (FAW) and Shanghai Automotive (SAIC). The new cooperation proves that OEMs still focus on partnerships, especially with Chinese companies, due to the highly volatile EV market in China (Gutmann, 2018).

BMW also cooperates with Brilliance in China for the development and production of EVs. At the end of 2017, BMW and Brilliance China Automotive Holdings Ltd. opened the new "High Voltage Battery Center" in Shenyang in China (BMW-Brilliance, 2017). The battery factory supplies the plant of the JV (BMW Brilliance Automotive) in Dadong, where PHVs will be produced for the Chinese market. According to recent news BMW and Brilliance have extended their JV contract until 2028. Currently the JV employs 16,000 people and accounts more than 350 local suppliers within their supply chain network (BMW Group, 2017).

Partnerships with technology companies

This year Nissan announced a strategic partnership agreement with the energy provider E.ON. Together, they want to pool their international expertise in the field of e-mobility and decentralized energy production and storage. Both companies are starting to launch pilot projects and joint customer offers soon. They will focus on charging EVs, integrating the charging infrastructure into the electricity grid, and decentralized energy generation and storage.

For both companies, the strategic partnership is an important move to establish e-mobility and increase customer acceptance. Nissan develops a comprehensive electric ecosystem for private customers and companies, while E.ON integrates electric mobility into the energy system. Merging these two core activities enables EV drivers to simply use and charge their vehicles anywhere. E.ON and Nissan have already begun with a practical implementation of the partnership in Europe. As a buyer of a Nissan Leaf, customers in Denmark benefit from a complete package, consisting of a charging station for their own home and an energy flat-rate in order to charge their EV battery. Additionally, Nissan and E.ON offer services which optimize customers energy consumption and total costs. Nissan customers can get solar and storage solutions for their home and can chose renewable energy solutions that also help to reduce energy costs (E.ON, 2018).

Another recent interesting cooperation happened between Jaguar and tech company Google.

In March 2018, Jaguar revealed the cooperation with Waymo which is the subsidiary of Google for self-driving cars. Jaguar will supply Waymo with "Jaguar I-Pace" electric cars, which will be equipped with Waymo's own technology. The Jaguar I-pace will be used for Waymo's driverless taxi services, which should start at the end of this year.

Over the next two years, Jaguar plans to deliver around 20,000 Jaguar I-Pace electric SUVs to Waymo. Both companies will work together to develop and build self-driving cars. According to recent news, both companies aim to cooperate on a long-term basis. In 2018, Waymo wants to take the first autonomous driving service onto the roads.

By means of this cooperation Jaguar is selling cars to Google on a B2B basis. This example shows the shift from traditional B2C into B2B business due to new ways of collaborations, which are caused by e-mobility (Heise, 2018). Additionally, it shows how OEMs are becoming a supplier for tech companies which are using EVs as hardware and are reproducing their digital concepts into the EV (Schaal, 2017).

The most important facts about the previous illustrated partnerships can be observed as followed:

- China as a key market for partnerships
- Rising partnerships between OEM and battery suppliers
- Joint Ventures become essential for partnerships
- Partnerships between energy provider
- Partnerships between Tech companies

According to a study from Deloitte about the e-mobility and how it shapes the automobile industry, shows how additional services such as financing solutions and mobility services will impact the revenue generation of car companies. Deloitte estimates that in 2025, car production only accounts for approx. 60% of the revenue of the industry and 40% come from new services. Depending on how EVs are developing it could also be that the share between production revenue and additional service revenue can account up to 50/50%.

This proves the urgency of OEMs to include added services into their business models.

New services have been already discussed within the previous chapter. They are mostly invented and offered by new entrants.

Mobility providers are offering innovative car sharing concepts.

Tech companies are providing customers a smarter car which is more connected to the external world.

Energy operators are developing smart and fast charging infrastructure for customers, which gives EV drivers the ability to recharge their EV fast and easy in urban city areas or also at home (Deloitte, 2017).

These are all value-adding services to the traditional car, which are all addressing customer preferences which have been addressed in Chap. 3.

If OEMs will not include added services into their business models, 50% of the automotive revenue would be generated by new entrants and competitors, according to Deloitte's e-mobility study (2017).

In order to avoid the losing revenue potential of, from a helicopter perspective, OEMs have to focus on two crucial issues.

First one deals with the question: How can we produce EVs in the most efficient and effective way?

Second issue deals with the question: How can we sell these? An additional question which can be derived from question 2: What do customers care about the most and how can we transform this need into an added service and this in our business model/value chain?

Regarding the current findings, basically five general strategies can be derived from these questions. Traditional OEMs have to:

1. Adapt their production process by generating new competences for battery and EV production.
2. Adapted after sales and financing solutions which address the requirements of EV buyers.
3. Add new mobility services.
4. Provide charging infrastructure.
5. Gain competencies from tech companies for future trends.

In the following course potential strategies for OEMs will be presented. Every strategy aims for the ability of traditional OEMs to compete in the future automobile industry, where cars will be electric, shared, and connected.

6.3.2 After Sales and Financing

Within this book, customer preferences in relation to electric cars have been analyzed. As in Chap. 3 customer preferences already have been mentioned, customer's acceptance for EVs relies heavily on the total cost of ownership. OEMs business models have to focus on, reducing the costs of total ownership, in order to increase customer acceptance. Innovative financing solutions and mobility concepts are addressing these preferences. Car sharing concepts are one possibility to spread the acquisition costs to a pool of drivers, which is one form of customer-oriented business model.

Electro-mobility can radically change the way people use automobiles. Instead of owning a vehicle, the flexible and integrated use of cars in connection with other ways of transport is gaining in importance.

Developing new business models for electro-mobility challenges OEMs (Siegfried, 2015). They must significantly reposition themselves as mobility service providers (Madan, 2018). Having a look on the scenario analysis of Chap. 4, the return of investing in new business models tends to be low due to the slowly raise of e-mobility at the beginning. However, with the outlook on 2030 and further, the investigations in such new concepts will definitely have a high return and have to be executed by OEMs, in order not to lose market share.

The battery costs of electric cars make up a significant proportion of the high acquisition costs compared with ICEs. In addition, customers connect uncertainties with batteries. Batteries may lose their storage capacity, as a result the range of the vehicles may decrease over time. To avoid additional acquisition costs and technical risks of batteries, battery leasing offers one potential business model. Manufacturers provide the battery which can be then borrowed by the customer for a user fee. By means of battery leasing, the performance of the battery is ensured to customers. They can replace their batteries as soon as the battery quality is reducing. Subsequently, battery leasing is reducing the initial cost of EVs because the price of the battery is not included in the price of the EV anymore. The cost of the battery is paid by a monthly or quarterly rate over the lifetime of the battery.

This model is an attractive feature for customers who are uncertain about the new technology of electric engines. Also OEMs can benefit from this model because they can generate additional revenue by charging fees on leasing rates. Additionally, OEMs can resell used batteries on the secondary market and generate additional earnings, if the used battery is still usable. However, the manufacturer carries the risk of battery capacity, because the OEM remains the legal owner of the battery (Deutsche Leasing, 2018).

As an alternative to the OEM, an external leasing company, battery manufacturer, or energy provider could provide the battery directly to the customer.

Mercedes-Benz and Renault are already offering battery leasing (Renault, 2018).

In addition to battery leasing, car leasing can be used by OEMs in order to reduce customers uncertainty toward new technologies. This business model is already well established in the automobile industry (Siegfried, 2013). In this case, a leasing contract is concluded between the customer and the manufacturer or the leasing company. The customer has to pay a monthly or quarterly leasing rate and in return receives the rights to use the vehicle. This is again an attractive model for customers, because he does not have to pay total acquisition costs at the beginning of the purchase and he can easily use the car for a few years and give it back to the manufacturer or the leasing company. The risk of the EV lies on the legal owner of the car and not on the customer (Deutsche Leasing Fleet GmbH, 2018).

Currently after sales and maintaining activities make up around 35% of the total revenue for OEMs and are responsible for a major part of the profit (Oliver Wyman, 2015). EVs consist of around 200 parts whereas ICEs consist of more than 1400 parts (Klesse, 2011). As a consequence, the relevance of maintaining and after sales will lose importance in the future. Due to the high share of digital and software components in EV, software updates have to be undertaken.

However, this will be made by over-air or cloud technology, which cuts out the traditional contact point between customer and manufacturer (Deutsche Leasing AG, 2018). Therefore, OEMs have designed full-service leasing for customers. Full-service contracts allow customers to benefit from repair and insurance services. This gives customers certainty toward the new technology, which is not easy for customers to understand in contrast to the mechanical ICE.

6.3.3 Mobility Provider

Urbanization and changes in the mobility preferences of customers led to new mobility concepts. Megatrends, such as using cars instead of owning cars, increased environmental awareness and the desire for more flexibility. Many manufacturers such as BMW and Daimler are currently striving to change from a traditional car manufacturer to a mobility provider.

Car sharing has gained huge acceptance in many customer groups and is likely to be one of the key mobility concepts of the future. In contrast to the traditional car rental, car sharing enables customers to use cars spontaneously and for a significantly shorter duration.

This business concept eliminates the barrier between customers and EVs, which is caused by the high purchase price of the EV. In addition, car sharing allows customers to drive new car models. For OEMs, this concept can be used for acquiring new customer groups. Customers who are 18–45 year old are no longer willing to buy their own car. These customer groups are also expensive for OEMs to acquire or even not acquirable So OEMs can use this model to expand their customer base. This concept automatically creates a direct contact point between manufacturer and customer. This is an ideal sales strategy to promote the car and the brand image of the manufacturer.

Regarding to Chap. 3, customers are more likely to buy EVs after they had first experience with driving an EV. By car sharing, users are initially brought together with EVs and can gain first experience, which increases the acceptance of new electric powertrains. Furthermore, the concept of car sharing is contributing to the development of EVs into a mass market because manufacturers can drop a large amount of EVs instantly. Simultaneously OEMs can collect user information of car sharing users and can therefore create new sale approaches based on the collected data (Bert et al., 2016).

Daimler already offers car sharing models like "Car2go" as well as BMW with "DriveNow."

While designing car sharing models, it is crucial for OEMs to define the reason for this concept. Car sharing should not only be implemented in order to use it as new sales strategy. It should be implemented in order to meet the changing customer preferences. Customer preferences are not primary to reduce CO_2 emissions, much more they want to travel from A to B flexibly in the most effect way. Efficient here does not mean only with low emissions but with low effort. Customers want mobility services they can use, without acquiring an expensive car and without having high costs. This is the primary requirement of customers when it comes to new mobility services. Once OEMs has figured out this, they will be able to design a future-oriented business model for shared economy (Helbig, 2018).

The aim of automobile manufacturers must be to achieve an integration of all transport possibilities, so that customers have the choice between efficient transport possibilities. OEMs have to build a mobility network for the customer. The customer then can decide which mobility service will most likely fulfill their preferences. It might be the case that the customer has to use public transportation in order to travel between two locations, but the customer needs an efficient service to get to the public transportation. OEMs can integrate public transportation into their mobility services and combine it with a car sharing service. Even if the vehicles only represent a part of the mobility chain and are no longer automatically the first choice for all distances, there are interesting business areas for automobile manufacturers.

The so-called mobility platforms can integrate the individual transport and offer the customer smooth mobility options from a single source.

It is important to create a mobility network that is as attractive as possible, so that the customer has a wide choice of transport options and can choose the best fitting option which suits their current situation. In order to create such a broad mobility network, it is important to consider almost every possible transportation possibility which is used by today's customers. Therefore, OEMs have to build partnerships with public transport, taxi companies, bicycle rental systems, free-floating and station-based car sharing operators (Etherington, 2018).

The transport for long distances can be covered by partnerships with railways, bus companies, carpools, and rental car companies. This enables customers to choose between numerous mobility options, which can be combined according to personal preferences. In addition, such an extensive service offers flexibility during a journey. In many cases, customers stuck in traffic jams while driving.

By means of mobility network platforms, customers can spontaneously change their option of transport and can be provided automatically with an alternative transport option which would reach to the targeted destination faster, more efficiently, or more environmentally friendly (Corwin et al., 2018).

In this car the customer would use the mobility platform and would confirm the choice of the offered alternative mobility option. In order to enable the customer the mobility option immediately, the booking and billing process will be executed by means of the platform. Customers should get their transport ticket for the chosen alternative transport option directly from the platform. So, customers do not have to make any effort except for choosing the most suitable alternative transport option, which is offered by the platform. In this way, the cost of a whole transport, which can consist of a car sharing trip, public transport, and bicycle rental, can be charged together in one single bill at the end of the trip.

This development must be supported by information and communication technologies so that customers can access the mobility platforms via smartphone apps. These apps provide an availability check by means of real-time information, access to timetables, costs, and routes of the transport option. Billing and booking should also be cashless and should be covered by the app (Goodall et al., 2017, pp. 114–119). All bookings can be paid by the customer at the end of the month. The customer does not have to make decisions on buying monthly bus tickets anymore. The app will automatically choose the cheapest transport ticket and the customer only has to pay the transport option he actually used.

In order to be able to operate car sharing models, OEMs have to ensure that EVs are used frequently by the customers around the clock. Only with such a high level of utilization, costs and maintenance of the vehicles can be compensated. Car sharing models have to meet the required level of utilization. Rural areas in contrast, have a significantly lower utilization level, which makes it challenging for OEMs to implement cars sharing models there. Therefore, OEMs have to find alternative models for these areas. One possible approach would be peer-to-peer car sharing models. In this model of sharing the rental of vehicles takes place between two private persons. This approach represents an innovative concept because it makes it relatively easy to reduce the number of vehicles on the road. Whereas car sharing does not immediately reduce the overall car fleet. Car sharing models only reduce the overall car fleet if it is highly utilized (Madan, 2018).

Shared economy is a game-changing trend which does not only impact the automobile industry. Airbnb started to expand shared economy around the world. In the automobile industry this trend has been shaped by Uber.

Sharing existing vehicles is much more efficient than adding high car pools for car sharing which are not fully utilized. Uber is a form of optimized car sharing models which have entered the everyday life of consumers in a very short period of time. Uber launched its mobility service app in 2010 and reached 30 million monthly active users in 2016 already (The Economist, 2016). In 2018, Uber has 75 million active users per month (Bhuiyan, 2018).

This shows that the need of innovative optimized car sharing models like Uber or Lyft meet a huge existing customer need. These models also provide a solution for crowded cities, as they have no space for new cars. This is criticized by car sharing concepts, which add up high car fleets to cities. Considering the high potential market for optimized car sharing concepts, OEMs should focus on the development of similar business concepts. However, the business concept of Uber is not legally accepted in numerous European countries such as Germany. Policy makers are suing Uber for taking away jobs of taxi drivers. Replacing taxi drivers in a county like Germany would mean a reduction of 40,000 jobs. This would mean a big disadvantage from national economy perspective. In April this year the European Court of Justice has decided that EU-nations and governments have the right to refuse the market entry of UBER in their country (Briegleb, 2018).

So policy makers still decelerate the growth of shared economy concepts although the customer need is tremendously high, also in European cities like London (3.5 million active users) (Titcomb, 2017).

OEMs should, therefore, develop similar models which are not that dependent on legally decisions. Alternatively OEMs can create similar optimized mobility concepts which are based on private car sharing models of Uber or Lyft. "Neighborhood sharing" enables local persons in a neighborhood to buy a vehicle together and use it jointly. Another alternative are networking platforms which connect travelers which use the similar routes. These travelers will than have the opportunity to share one car with travelers who have the same route. This concept of shared economy contributes to the solution of relieving crowded cities without being dependent on policy makers like Uber's mobility concept.

Mohseni (2018) explains the relevancy of mobility services for the future.

He thinks that partnerships with municipalities of large cities are becoming increasingly important. He mentions cities such as Dubai or Singapore which will in future rely entirely on autonomous vehicles and digitized cities. Continental supports the development of smart cities heavily. The aim is to make these cities safer, more environmentally friendly, and more efficient in order to meet the mobility needs of the residents. Mohseni also tells that Continental is expanding with its first integrated solutions for areas such as sharing, parking, automation, fleet management, intelligent street lighting and safe road intersections.

A city without accidents or pollution is the key purpose of the so-called smart cities. The trend of smart cities includes new concepts and new forms of mobility, which make big cities a future place to be. A smart city consist of autonomous EVs, car sharing models, self-parking cars, connected cars, and traffic lights which manage a smooth and efficient traffic. It is all about how to make urban cities more efficient although population is growing consistently. Smart cities should be efficient in their consumption of space and energy while reducing the risk of accidents. Additionally, a huge amount of data will be accessible, due to the concept of smart cities because they are based on digitalization and technology (Becker, 2017; Grimming, 2015).

In the following, five smart cities will be listed in order to understand the concept of smart cities. These cities are heavily contributing to the development of the future mobility (Mann, 2016).

Tokyo is planning an innovative project for their Olympic summer games in 2020. During the event policy makers and partnerships have revealed that self-driving vehicles will be in charge of transport visitors around the city. According to current news Asian OEMs as Honda and Nissan as well as robotic and Internet companies are heading to implement more than thousands of self-driving vehicles.

As Mr. Mohseni explains in the interview for this book, Singapore will probably become a future mobility city.

In 2016, Singapore was the first city which has driverless taxis on their roads. This pilot project aims to have a fully autonomous driving taxi fleet in Singapore. This project is also strongly supported by policy makers.

Boston is participating on the purpose of future mobility. Boston started a city project which should support the successful integration of a driverless future. Therefore, Boston allows on-street testing of autonomous cars since 2016.

In Amsterdam, a semi-autonomous Mercedes bus is transporting travelers between the airport of Amsterdam to the city of Amsterdam. This pilot project which have been launched in 2016, aims to set the basis for the public transport of the future.

A very interesting smart city concept can be seen in Pittsburgh where self-driving UBER taxis are introduced since 2016.

Pittsburgh accepted to let UBER use their streets for, developing and test their technology. Uber users in Pittsburgh have the option to select a "self-driving car" for their transportation. Uber started to test 14 self-driving Uber cars in the city location.

In London, a project has been settled which specializes on the use and acceptance of autonomous vehicles. This project aims to contribute to the development of driverless cars in the world. In this project people from London can register to use fully electric autonomous cars. Participants will be interviewed and monitored during this project. This enables an enormous data collection and creates significant value for the implementation of EVs.

6.3.4 Technology Companies

Besides mobility services like car sharing, technology will shape the automotive value chain and the whole industry tremendously. This chapter is referring to the technology point of view. Digitalization, connectivity, artificial intelligence, and autonomous driving are all technology based trends, which will be a central element for OEMs and suppliers.

From the expert interview it can be observed that also from the supplier point of view, technology companies will play a key role in the future automotive industry.

Mr. Mohseni, head of business development of Continental, highlights the importance of partnerships with technology companies and suppliers. The future electric car will be digital, autonomous, smart, and connected.

Mr. Mohseni also points out that suppliers and car manufacturers must invest in new market entrants or even have to take over these companies completely by M & A transactions. Many start-ups already offer solutions for the future today. He also mentions the Israeli company Argus Cyber Security, which Continental acquired in November 2017.

There are four key trends which are forcing OEMs and suppliers to enlarge the automotive value chain by the field of "technology."

These trend are autonomous driving (self-driving), connectivity (connected), artificial intelligence (smart and safe), and digitalization (user-oriented). Almost every tech company and start-up which is entering the automobile industry is focusing on one of these trends.

The head of Continental also says that OEMs and tier-1 suppliers will pass on their traditional activities. The future of large OEMs lies mainly in engineering and know-how, rather than in the production of components.

Engineering and know-how can only be acquired by partnerships with technology companies or start-ups.

As the interview with Continental showed, the future electric car will have totally different distinguishing feature than ICE's.

He adds that the future EVs will distinguish itself by driver assistance systems, holistic networking solutions, safety technologies which save lives. The vehicle of the future will be all-time connected, user-friendly, comfortable, and intelligent. Differentiation characteristics of the vehicles will be above all individual mobility offers. One such future technology is augmented reality applications based on our "AR head-up display technology." As a result OEMs and suppliers have to focus on collaborations and acquisitions which focus on these areas.

According to Mr. Mohseni, these fields will be in a strong focus of suppliers.

Manufacturers, in contrast, have their own visions of future differentiation features. He is underlining this, by giving the example of Daimler. Daimler's Mercedes-Benz user experience in the connectivity and intuitive operation which is managed by smart voice control systems, allow a user-friendly experience which have never been felt before.

All manufacturers and suppliers have their own projects and ideas which they offer and monitor in order to see which one will prevail.

The head of business development of the leading automobile supplier also says that artificial intelligence is indispensable for autonomous driving and the future. That is why Continental is investing heavily in partnerships such as with NVIDIA. Together, they want to provide a complete, AI-based solution for self-driving vehicles. In 2021, Continental will launch the autonomous vehicle system based on the "Nvidia AI" vehicle computer and sensors. The partnership will enable the development of AI computer systems ranging from automated level 2 functions to complete level 5 self-propelled functions where the vehicle has neither s steering wheel nor pedals.

He also reveals that Continental entered into a further cooperation with the German Research Center for Artificial Intelligence (Deutsche Forschungszentrum für Künstliche Intelligenz) in June 2018. Continental benefits at all levels of the company, for example, in knowledge management, material flow, and in software and product development. AI-based software and automatic learning are the basis for the future of automated and autonomous driving. Self-learning systems for smart vehicles, which understand the gestures and intentions of pedestrians or other road users, help the environment to make the future more safety and more enjoyable.

6.3.5 Infrastructure

The provision of charging infrastructure and power supply for electric cars is a totally new business for OEMs which cannot be covered by the existing competencies of the automotive industry.

The development of the charging infrastructure is a crucial factor for OEMs, since EVs can only be used if a charging station network exists. Energy and infrastructure providers are important parties at the end of the new automobile value chain, which lay the basis for using EVs.

Charging infrastructure concepts are often tested in pilot projects for a specific time period. These projects require the cooperation between OEMs and energy providers.

EnBW Energie Baden-Württemberg and Daimler launched several pilot projects such as "iZeus" or "CROME" which aim to develop a model region for locally emission-free EVs (EnBW, 2018). Another example is the alliance between petrol station provider Shell, BMW, Daimler, and VW. The collaboration between these OEMs and Shell is called "Ionity" and aims to build up to 400 fast charging stations in Europe, which can charge EVs within 5–8 min (Hubik, 2017).

OEMs can build partnerships with charging station providers. They can provide private charging stations for their customers. Wall boxes are usually integrated on the wall of the private household and link the EV to the electricity grid. BMW has already collaborated with charging infrastructure provider Schneider Electric to organize the installation and maintenance of such private wall boxes (BMW, 2013). Due to such collaborations OEMs can meet customer preferences. They reduce customers uncertainty at the time of the purchase, and reduce doubts regarding charging the EV. This increases the willingness of the buyers and therefore increases EV sales. Additionally, private stations would benefit from second life capacity of used batteries.

After the standard operating life of a battery, they are no longer useful for EVs. However, they still have a high storage capacity and can be used in private households. Functioning cells of the EV battery can be assembled into a flexible storage system for the private household. These flexible energy storage batteries can temporarily store the energy of the in-house solar energy systems. For the case that the solar system does not supply sufficient power, the energy storage can be used so that electricity can be stored cost-effectively at for a period of time (Reid & Julve, 2016, pp. 20–25). Based on this concept future innovations allow households to use smart grids, which are relieving overloaded electricity grids and are allocating electricity efficiently and cheaper for users (Siemens, 2018).

Private charging stations are suitable for rural areas. In city areas, in contrast, people cannot use private charging stations because of a lack of space. Therefore, it will be focused on public charging stations in city areas. This comes with an increasing complexity and more challenges. The required public infrastructure is currently under construction and has to struggle to reduce charging duration. The charging duration in city areas has to be significantly lower than in private households. Long charging stations in cities would mean that EVs have to stay for a longer time at stations. As a consequence more stations are required to meet the charging demand of EVs. Due to a shortage of space in cities, it is difficult to build more stations. Therefore, it has to be focused on building more efficient stations which charge in a faster time period (Sia Partners, 2016).

The provision of private charging stations is a potential business model for OEMs, however public charging stations seem to be to cost intensive and not lucrative for OEMs.

Currently the public charging network has been largely built up by automotive and energy providers, in form of pilot projects (Mühlon, 2017). OEMs have stated that investing in infrastructure is not a strategic focus due to low returns, because the number of EVs is still relatively low. According to the interview with Head of Business Development of Continental, Mohseni (2018), the majority of OEMs are no longer focusing on building the required charging infrastructure and do not primarily plan to develop new business models in this area. The chief executive officer of Volkswagen Group CEO said few years ago that it is not the task and business model of the automotive industry to set up charging stations.

Therefore it is unlikely to see that OEMs will be actively developing large charging infrastructure projects in the future.

Nevertheless, there are and will be pilot projects and alliances between manufacturers and energy providers for building charging infrastructure. The responsibility and development of the charging infrastructure will mainly come from infrastructure providers and energy suppliers, while manufacturers will focus more on the topics mentioned above, such as collaborations with suppliers, financing, mobility services, new mobility concepts, and technological development.

References

Accenture. (2014). *The electric vehicle challenge*. Retrieved September 20, 2021, from Accenture Website: https://www.accenture.com/in-en/~/media/Accenture/Conversion-Assets/DotCom/Documents/Global/PDF/Industries_15/Accenture-Electric-Vehicle-Challenge.pdf

Becker, J. (2017, July 7). *Stressfrei in die Zukunft*. Retrieved September 20, 2021, from Süddeutsche Zeitung Online: http://www.sueddeutsche.de/auto/smart-city-stressfrei-in-die-zukunft-1.3572959

Bert, J., Collie, B., Gerrits, M., & Xu, G. (2016, February 23). *What's ahead of car Sharing*. Retrieved September 20, 2021, from Boston Consulting Group: https://www.bcg.com/de-de/publications/2016/automotive-whats-ahead-car-sharing-new-mobility-its-impact-vehicle-sales.aspx

Bhuiyan, J. (2018, January 5). *Uber powered four billion rides in 2017. It wants to do more—and cheaper—in 2018*. Retrieved September 20, 2021, from Recode: https://www.recode.net/2018/1/5/16854714/uber-four-billion-rides-coo-barney-harford-2018-cut-costs-customer-service

BMW. (2013, January 18). *BMW i besiegelt globale Partnerschaft für kundenfreundliche Elektromobilität*. Retrieved September 20, 2021, from BMW Group: https://www.press.bmwgroup.com/deutschland/article/detail/T0136262DE/bmw-i-besiegelt-globale-partnerschaft-fuer-kundenfreundliche-elektromobilitaet?language=de

BMW Group. (2017, October 24). *BMW brilliance automotive eröffnet Batteriefabrik in Shenyang*. Retrieved September 20, 2021, from BMW Group Website: https://www.press.bmwgroup.com/deutschland/article/detail/T0275323DE/bmw-brilliance-automotive-eroeffnet-batteriefabrik-in-shenyang?language=de

BMW-Brilliance. (2017, October 24). *BMW Brilliance Automotive Opens High Voltage Battery Center in Shenyang*. Retrieved September 20, 2021, from BMW-Brilliance Website: http://www.bmw-brilliance.cn/cn/en/news/news/2017-10-24.html

Boston Consulting Group. (2018). *Joint ventures alliances*. Retrieved September 20, 2021, from Boston Consulting Group Website: https://www.bcg.com/de-de/capabilities/mergers-acquisitions-divestitures/joint-ventures-alliances.aspx

Briegleb, V. (2018, April 10). *EuGH: EU-Staaten dürfen Uber-Dienste verbieten*. Retrieved September 20, 2021, from Heise: https://www.heise.de/newsticker/meldung/EuGH-EU-Staaten-duerfen-Uber-Dienste-verbieten-4014689.html

Brühl, T. (2011). *Thesen, Hypothesen, Annahmen*. Retrieved September 20, 2021, from Goethe Universität Frankfurt am Main: http://www.fb03.uni-frankfurt.de/46036826/thesen_hypothesen_annahmen.pdf

BYD. (2018). *Company profile*. Retrieved September 20, 2021, from BYD Website: http://bydeurope.com/company/profile.php

Continental. (2018, March 3). *Continental und CITC gründen Joint Venture zur Produktion von 48-Volt-Batteriesystemen*. Retrieved September 20, 2021, from Continental Website: https://www.continental-corporation.com/de/presse/pressemitteilungen/2018-03-07-jv-citc-123416

Corwin, S., Hood, J., Dinamani, A., Skowron, J., & Pankratz, D. (2018, March 9). *Cities explore digital mobility platforms.* Retrieved September 20, 2021, from Deloitte: https://www2.deloitte.com/insights/us/en/focus/future-of-mobility/urban-transport-mobility-platforms.html

Daimler AG. (2014, October 21). *Daimler strukturiert Zusammenarbeit mit Tesla neu.* Retrieved September 20, 2021, from Daimler Website: http://media.daimler.com/marsMediaSite/de/instance/ko/Daimler-strukturiert-Zusammenarbeit-mit-Tesla-neu.xhtml?oid=9920370

Deloitte. (2017). *The future of the automotive value chain.* Retrieved September 20, 2021, from Deloitte Website: https://www2.deloitte.com/content/dam/Deloitte/de/Documents/consumer-business/Deloitte_Sonderbeilage%20Automobilwoche2017_safe.pdf

Deutsche Leasing AG, O. M. (2018, May 11). *Operations manager.* Deutsche Leasing AG. (K. Ciftci, Interviewer). Bad Homburg von der Höhe, Hessen.

Deutsche Leasing Fleet GmbH, M. (2018, May 11). *Manager* (K. Ciftci, Interviewer).

Deutsche Leasing, A. M. (2018, May 14). *Account manager* (K. Ciftci, Interviewer) Bad Homburg von der Höhe, Hessen.

E.ON. (2018, March 6). *eMobility: E.ON und Nissan vereinbaren strategische Partnerschaft.* Retrieved September 20, 2021, from E.ON Website: https://www.eon.com/de/ueber-uns/presse/pressemitteilungen/2018/emobility-eon-und-nissan-vereinbaren-strategische-partnerschaft.html

EnBW. (2018). *E-Mobilitätsprojekte.* Retrieved September 20, 2021, from EnBW Website: https://www.enbw.com/privatkunden/energie-und-zukunft/e-mobilitaet/projekte/abgeschlossene-projekte.html

Etherington. (2018, January 9). *Ford and Autonomic are building a smart city cloud platform.* Retrieved September 20, 2021, from Tech Crunch: https://techcrunch.com/2018/01/09/ford-and-autonomic-are-building-a-smart-city-cloud-platform/?guccounter=1

Fuchslocher, G. (2018, March 26). *Daimler und BYD stellen den DENZA 500 vor.* Retrieved September 20, 2021, from Automobil-Produktion: https://www.automobil-produktion.de/hersteller/daimler-und-byd-stellen-den-denza-500-vor-302.html

Goodall, W., Fishman, T., Bornstein, J., & Bonthron, B. (2017). The rise of mobility as a service. In Deloitt (Ed.), *Deloitte Review* (20, pp. 111–130).

Grimming, R. (2015, November 18). *Audi sucht in der "Smart City" die Zukunft des Autos.* Retrieved September 20, 2021, from Heise: https://www.heise.de/newsticker/meldung/Audi-sucht-in-der-Smart-City-die-Zukunft-des-Autos-2924337.html

Günnel, T. (2014, March 3). *Continental übernimmt Zytek automotive.* Retrieved September 20, 2021, from Automobil-Industrie website: https://www.automobil-industrie.vogel.de/continental-uebernimmt-zytek-automotive-a-441494/

Gutmann, T. (2018, June 5). *E-Autos: Volkswagen plant drei neue Werke in China.* Retrieved September 20, 2021, from Ostexperte: https://ostexperte.de/volkswagen-china/

Hajek, S., Seiwert, M., Deuber, L., & Fritz, M. (2017, June 17). *Akkus für Elektroautos.* Retrieved September 20, 2021, from Wirtschafts Woche Website: https://www.wiwo.de/technologie/mobilitaet/akkus-fuer-elektroautos-jahrzehntelang-geschlafen/19905636-2.html

Heise. (2018, March 28). *Autonome Autos: Waymo baut Elektroautos von Jaguar zu Robotaxis um.* Retrieved September 20, 2021, from Heise: https://www.heise.de/newsticker/meldung/Autonome-Autos-Waymo-baut-Elektroautos-von-Jaguar-zu-Robotaxis-um-4007196.html

Helbig, K. (2018, March 17). *ar share, public transport and walking: better ways to get from A to B.* Retrieved September 20, 2021, from The Guardian: https://www.theguardian.com/lifeandstyle/2018/mar/18/car-share-public-transport-and-walking-better-ways-to-get-from-a-to-b

Hubik, F. (2017, November 27). *Shell tritt der Schnelllader-Allianz bei.* Retrieved September 20, 2021, from Handelsblatt: https://www.handelsblatt.com/unternehmen/industrie/auto-von-morgen/oelriese-kooperiert-mit-deutschen-autobauern-shell-tritt-der-schnelllader-allianz-bei/20634184.html?ticket=ST-1448210-bSlRn1AbM5b7lMxZwdJO-ap4

Huke, J./Siegfried, P. (2021): Finance methods in the automotive sector - business agility in the age of digital disruption, International Journal of Automotive Science and Technology, 5, 3, 281–288, https://doi.org/10.30939/ijastech..955551

Kamp, M. (2010, March 23). *Womit Chinas Autohersteller BYD wirklich auftrumpft*. Retrieved September 20, 2021, from Wirtschafts Woche Website: https://www.wiwo.de/unternehmen/auto/elektroautos-womit-chinas-autohersteller-byd-wirklich-auftrumpft-seite-2/5143950-2.html

Koch, W. (2006). *Wertschöpfungskette Automobile*. Gabler Verlag.oliver.

Klesse, H. (2011, September 14). *Autohersteller büßen Macht ein*. Retrieved September 20, 2021, from Wirtschafts Woche: https://www.wiwo.de/technologie/mobilitaet/elektroautos-autohersteller-buessen-macht-ein-seite-2/5212332-2.html

Krust, M. (2012, March 30). *Daimler und BYD bringen E-Auto unter Marke Denza*. Retrieved September 20, 2021, from Automobilwoche: https://www.automobilwoche.de/article/20120330/NACHRICHTEN/120339991/1318/article/99999999/REPOSITORY/110729976/agbdatenschutz

Lebedeva, N., Persio, F., & Boon-Brett, L. (2016). *Lithium ion battery value chain and related opportunities for Europe*. European Comission.

Liu, J., & Brody, P. (2014). *Is collaboration the new innovation* (H. B. Review, Ed.). Retrieved September 20, 2021, from Ernst & Young Website: https://www.ey.com/Publication/vwLUAssets/ey-is-collaboration-the-new-innovation/$FILE/ey-is-collaboration-the-new-innovation.pdf

Madan, A. (2018, May 7). *Partner, Consulting, KPMG AG Wirtschaftsprüfungsgesellschaft* (K. Ciftci, Interviewer). Frankfurt am Main.

Mann, K. (2016, October 14). *10 cities at the forefront of automated driving*. Retrieved September 20, 2021, from 2025 AD - The Automoted Driving Community: https://www.2025ad.com/latest/driverless-cities/

Mohseni, A. (2018, June 4). *Head of business development, continental AG* (K. Ciftci, Interviewer, & C. Kaan, Translator). Escborn, Hessen.

Mühlon, F. (2017), October 10). *Umbruch in der Automobilbranche durch Elektromobilität*. Retrieved September 20, 2021, from ABB Website: https://www.abb

Oliver Wyman. (2015, September 15). System profit. Retrieved September 20, 2021, from Oliver Wyman website press release: http://www.oliverwyman.com/content/dam/oliver-wyman/global/en/files/who-we-are/press-releases/OliverWyman_Graphics_System_Profit_2035-Reinvent_Sales_EN_15092015_final.pdf

Özelci, M. (2018, May 3). *CEO and owner - Audi car dealer in Frankfurt* (K. Ciftci, Interviewer).

Reid, G., & Julve, J. (2016). *Second life-batteries as flexible storage for renewable energies*. Bundesverband Erneuerbare Energie; Hannover Messe.

Renault. (2018). *Renault ZOE - Batteriemiete*. Retrieved September 20, 2021, from Renault Website: https://www.renault.de/modellpalette/renault-modelluebersicht/zoe/Batteriemiete.html

Reuters. (2017, May 22). *Chinese regulator approves VW-JAC Motor electric car venture*. Retrieved September 20, 2021, from Reuters: https://www.reuters.com/article/us-jac-volkswagen/chinese-regulator-approves-vw-jac-motor-electric-car-venture-idUSKBN18I15I

Roland Berger. (2017, February). *Automotive metal components for car oddies and chassis*. Retrieved September 20, 2021, from Roland Berger website: https://www.rolandberger.com/publications/publication_pdf/roland_berger_global_automotive_stamping_study_e_20170210.pdf

Rother, F., Rees, J., & Klesse, H. (2010, May 6). *Die Chancen der Zuliefer beim Elektroauto*. Retrieved September 20, 2021, from Wirtschafts Woche Website: https://www.wiwo.de/unternehmen/auto/automobilbranche-die-chancen-der-zulieferer-beim-elektroauto-seite-2/5641668-2.html

Schaal, S. (2017, January 11). *Autokonzerne und Tech-Giganten.* Retrieved September 20, 2021, from Wirtschafts Woche: https://www.wiwo.de/unternehmen/auto/autokonzerne-und-tech-giganten-wer-macht-es-mit-wem-und-warum/19234002.html

Sia Partners. (2016, September 27). *Public charging infrastructures.* Retrieved September 20, 2021, from Sia Partners: http://energy.sia-partners.com/20160927/public-charging-infrastructures-essential-mass-adoption-electric-vehicles

Siegfried, P. (2013). *The service engineering concept for business* (pp. 173–187). Entrepreneurship-conference University of Lisboa/PRT. ISBN: 978-989-732-207-5.

Siegfried, P. (2014). *Analysis of the service research studies in the German research field, performance measurement and management.* Publishing House of Wroclaw University of Economics. ISBN: 978-83-7695-473-8 (Band 345, pp. 94–104).

Siegfried, P. (2015). Die Unternehmenserfolgsfaktoren und deren kausale Zusam-menhänge. *Zeitschrift Ideen- und Innovationsmanagement, Deutsches Institut für Betriebswirtschaft GmbH/Erich Schmidt Verlag.* ISSN 2198-3143, 131–137.

Siemens. (2018). *Connecting Grids.* Retrieved June 11, 2018, from Siemens Website: https://www.siemens.com/global/en/home/products/energy/topics/connecting-grids.html

Stricker, K., Matthies, G., & Tsang, R. (2011). *Vom Automobilbauer zum Mobilitätsdienstleister.* Retrieved September 20, 2021, from Bain & Company Website: http://www.bain.de/Images/Bain_brief_vom_automobilbauer_zum_mobilitatsdienstleister.pdf

Talgeri, V. (2014). *Outsourcing der Fertigung in den asiatischen Wirtschaftsraum: Risiken und Grenzen.* Diplomica Verlag.

The Economist. (2016, September 3). *Uber: From zero to seventy (billion).* Retrieved September 20, 2021, from The Economist: https://www.economist.com/briefing/2016/09/03/from-zero-to-seventy-billion

Titcomb, J. (2017, September 26). *Who really are Uber's 40,000 drivers and 3.5m users in London?.* Retrieved September 20, 2021, from The Telegraph: https://www.telegraph.co.uk/technology/2017/09/26/really-ubers-40000-drivers-35m-users-london/

Wimmelbücker, S. (2017, September 13). *Batterien für Elektroautos.* Retrieved September 20, 2021, from Automobilwoche Website: https://www.automobilwoche.de/article/20171213/NACHRICHTEN/171219953/batterien-fuer-elektroautos-toyota-und-panasonic-denken-ueber-gemeinsame-entwicklung-nach

Chapter 7
Conclusion of the Potential Impact of E-Mobility on the Automotive Value Chain

Due to the present development of e-mobility and technological trends, this book tried to investigate the impact of e-mobility and associated trends such as shared economy on the automobile industry. The book was designed to answer the question of how e-mobility is impacting the automotive value chain of traditional manufacturers, thereby contributing to the adaption to the new era of mobility.

The potential impact of e-mobility on the automotive value chain is broad. E-mobility is very likely replacing the traditional internal combustion engine in the future. The global market share of electric vehicles is increasingly becoming larger and is likely to surpass the market share of internal combustion engines between 2030 and 2040. The book showed which factors are driving this development. Customer preferences, technological innovations, and charging infrastructure are main drivers for the progress of e-mobility. Due to changing customer preferences, different ownership models which include electric vehicles, will gain huge acceptance. Additionally, technological innovations will reduce production costs of batteries and also the maintenance costs of electric vehicles. Consequently, a significant reduction in total cost of ownership can be observed, which has strong potential to accelerate the progress of e-mobility. The third driver "charging infrastructure" is contributed by cross-industry alliances, which are pushing the development of charging stations forward by high investments and pilot projects.

The book has underlined, that e-mobility cannot be seen isolated. E-mobility is the initial trigger for numerous mega trends. Trends like artificial intelligence, autonomous driving, and shared mobility are hitting the market simultaneously. Consequently, car manufacturers are competing with cross-industry forces. Technology companies, start-ups, mobility providers, and energy providers are entering the global automobile industry and are implementing new services and products to the market and, thus, force car manufacturers to change their traditional automotive value chain.

The upstream changes of the value chain are consisting of forming new supplier networks and increasing the degree of vertical integration.

K. Y. Ciftci et al., *The Potential Impact of E-Mobility on the Automotive Value Chain*, SpringerBriefs in Business, https://doi.org/10.1007/978-3-030-95599-1_7

Based on the new trends and new entrants, car manufacturers are extending their supplier network. Manufacturers are significantly increasing the depth of added value by entering cross-industry partnerships and changing their supplier network. The production of batteries and the development of new technologies like self-driving cars are pushing manufacturers toward vertical integration. The two main challenges in this field for manufacturers will be the selection of the supplier and the degree of outsourcing.

Downstream changes in the automotive value chain are including after sales and financing, mobility concepts and services, and charging infrastructure.

Manufacturers will add new financing and after sales services to their value chain, in order to reduce the total cost of ownership for customers, which is a significant factor for customers' buying decision. Besides EV leasing, battery leasing offers great potential for manufacturers and their customers.

Furthermore, the link to new mobility concepts, which change the principal idea of getting from A to B, was outlined. Car sharing is changing the traditional purpose of cars, which used to be a status symbol. Car sharing is transforming this perspective and enables people to share cars and use them flexibly. The main challenge for manufacturers is the development of a mobility concept, which includes all transport possibilities in order to fully meet the customer's requirements. People do not only put cars into their decisions for personnel transportation. Now, people have much broader attractive choices for traveling. Therefore, manufacturers also have to consider public transport, bicycle sharing, and car sharing when they are designing new mobility concepts and services. Manufacturers can provide car sharing platforms and offer mobility platforms. Car sharing platforms like Car2go and DriveNow are already implemented by traditional manufacturers. Mobility platforms still have to improve in order to gain a high market share. Innovative mobility platforms in the future will contain a broad view on different transport possibilities, artificial intelligence, and cashless payment, which can be used via smartphone app.

Additionally, the cultural aspect has been addressed briefly within this book. Car sharing models and new mobility concepts are changing behavioral thinking of consumers, which causes a cultural shift. Mobility concepts like Uber and Lyft get people into a "strangers" car to let them drive to a specific destination. Cultural behavior is hard to change and happens differently in every culture. With new mobility concepts, manufacturers are forcing people to change their cultural mindset.

A following research question could be "How can car manufacturers manage the cultural change of new mobility concepts?". This research would assume that cultural change drives the acceptance of EV. Based on this assumption, manufacturers have to implement a strategy to manage cultural change, in order to increase their revenue as mobility and EV provider.

The book further highlights, that automobile manufacturers are increasingly moving away from their traditional core business. They are changing from a car manufacturer to a mobility provider. The automotive market is dissolving its traditional boundaries and is merging with other industries. This creates room for manufacturers to reposition themselves in the new electric automobile industry.

Accordingly, an important finding is the potential fields for market repositioning. Manufacturers can focus on technical development by forming deep supplier networks or they can focus on downstream areas and aim to become a global mobility provider, or both. However, this decision represents an important strategic decision for manufacturers and has to be made individually for each manufacturer.

In a separated research, this strategic decision could be a subject of discussion. Strategic decisions are highly dependent on external market circumstances. Therefore, the researcher could choose a specific manufacturer. Based on the analysis of internal and external factors of this manufacturer, the researcher could give a potential strategy for repositioning in the future automotive market.

Overall, it can be said, that, car suppliers will take over a high share of the automotive value creation. Manufacturers are less and less focusing on upstream activities of the value chain. The electric vehicle will contribute to the shift of production and R&D toward suppliers. The strongest distinguishing feature of traditional cars is the engine. Due to e-mobility, new distinguishing features are replacing the traditional differentiation feature (Mohseni, 2018). Nevertheless, it is questionable if suppliers will take over more competencies of the manufacturer and will become "the new automobile manufacturers."

Reference

Mohseni, A. (2018, June 4). Head of business development, continental AG. (K. Ciftci, Interviewer, & C. Kaan, Translator) Eschborn, Hessen.